Hox Genes: Studies from the 20th to the 21st Century

ADVANCES IN EXPERIMENTAL MEDICINE AND BIOLOGY

Recent Volumes in this Series

Volume 681
MELANOCORTINS: MULTIPLE ACTIONS AND THERAPEUTIC POTENTIAL
Edited by Anna Catania

Volume 682
MUSCLE BIOPHYSICS: FROM MOLECULES TO CELLS
Edited by Dilson E. Rassier

Volume 683
INSECT NICOTINIC ACETYLCHOLINE RECEPTORS
Edited by Steeve Hervé Thany

Volume 684
MEMORY T CELLS
Edited by Maurizio Zanetti and Stephen P. Schoenberger

Volume 685
DISEASES OF DNA REPAIR
Edited by Shamim I. Ahmad

Volume 686
RARE DISEASES EPIDEMIOLOGY
Edited by Manuel Posada de la Paz and Stephen C. Groft

Volume 687
BCL-2 PROTEIN FAMILY: ESSENTIAL REGULATORS OF CELL DETH
Edited by Claudio Hetz

Volume 688
SPHINGOLIPIDS AS SIGNALING AND REGULATORY MOLECULES
Edited by Charles Chalfant and Maurizio Del Poeta

Volume 689
HOX GENES: STUDIES FROM THE 20TH TO THE 21ST CENTURY
Edited by Jean S. Deutsch

Hox Genes
Studies from the 20th to the 21st Century

Edited by

Jean S. Deutsch, PhD

Developmental Biology, Pierre and Marie Curie University

Paris, France

Springer Science+Business Media, LLC

Landes Bioscience

Springer Science+Business Media, LLC
Landes Bioscience

Printed in the USA.

Springer Science+Business Media, LLC, 233 Spring Street, New York, New York 10013, USA
http://www.springer.com

Please address all inquiries to the publishers:
Landes Bioscience, 1002 West Avenue, Austin, Texas 78701, USA
Phone: 512/ 637 6050; FAX: 512/ 637 6079
http://www.landesbioscience.com

The chapters in this book are available in the Madame Curie Bioscience Database.
http://www.landesbioscience.com/curie

Hox Genes: Studies from the 20th to the 21st Century, edited by Jean S. Deutsch. Landes Bioscience/ Springer Science+Business Media, LLC dual imprint / Springer series: Advances in Experimental Medicine and Biology.

ISBN: 978-1-4419-6672-8

While the authors, editors and publisher believe that drug selection and dosage and the specifications and usage of equipment and devices, as set forth in this book, are in accord with current recommendations and practice at the time of publication, they make no warranty, expressed or implied, with respect to material described in this book. In view of the ongoing research, equipment development, changes in governmental regulations and the rapid accumulation of information relating to the biomedical sciences, the reader is urged to carefully review and evaluate the information provided herein.

Library of Congress Cataloging-in-Publication Data

Library of Congress Cataloging-in-Publication Data

Hox genes : studies from the 20th to the 21st century / edited by Jean S. Deutsch.
p. ; cm. -- (Advances in experimental medicine and biology ; v. 689)
Includes bibliographical references and index.
ISBN 978-1-4419-6672-8
1. Homeobox genes. I. Deutsch, Jean S., 1942- II. Series: Advances in experimental medicine and biology ; v. 689.
[DNLM: 1. Genes, Homeobox. 2. Evolution, Molecular. 3. Gene Expression Regulation. W1 AD559 v.689 2010 / QU 470 H8679 2010]
QH447.8.H65H69 2010
571.8'61--dc22

2010015817

FOREWORD

In his 1894 book, *Materials for the Study of Variation*, William Bateson coined the term Homoeosis with the following prose:

> *The case of the modification of the antenna of an insect into a foot, of the eye of a Crustacean into an antenna, of a petal into a stamen, and the like, are examples of the same kind.*
>
> *It is desirable and indeed necessary that such Variations, which consist in the assumption by one member of a Meristic series, of the form or characters proper to other members of the series, should be recognized as constituting a distinct group of phenomena.*
>
> *...I therefore propose...the term HOMOEOSIS...; for the essential phenomenon is not that there has merely been a change, but that something has been changed into the likeness of something else.*

The book was intended as a listing of the kinds of naturally occurring variation that could act as a substrate for the evolutionary process and Bateson took his examples from collections, both private and in museums, of materials displaying morphological oddities. Interestingly the person who also coined the term "Genetics" proffered little in the way of speculation on the possible genetic underpinnings of these oddities. It wasn't until the early part of the next century that these changes in meristic series were shown to be heritable.

As was the case for Bateson, the homeotic or Hox genes were first recognized by virtue of their striking mutant phenotypes in the fly *Drosophila melanogaster*. The seminal work of E.B. Lewis enlarged on these early discoveries and mapped a set of these mutations into a complex on the third chromosome that affected the segmental identity of the posterior thorax and the abdomen. Subsequently a separate complex controlling the identities of the anterior thorax and head was discovered. The anterior and posterior acting complexes were named after their most prominent member loci: *Antennapedia* (ANT-C) and *bithorax* (BX-C) respectively. The genetic mapping of lesions within each complex and the segments affected by those lesions revealed the fact that the left to right order of the homeotic loci within each complex and the domains of affect of each locus were colinear. The subsequent molecular characterization of both complexes demonstrated that the expression domains of the genes were entirely consistent with the morphological results. The subsequent molecular characterization of the two complexes produced two other striking findings:

- The resident loci encoded transcription factors that contained a highly conserved motif termed the homeodomain
- The ANT-C contained loci that while encoding the homeodomain did not act as canonical Hox genes.

With molecular probes in hand it became possible to search for similar genes outside of the fly. It soon became apparent that not only are the Hox genes highly conserved, so is their residence in a complex and the rule of colinearity, with both characteristics extending to most animal phyla. Within the arthropods one can see excellent examples of this conservation despite rather divergent embryonic morphologies and types of embryonic development.

In the early days the discovery of the molecular character and ubiquity of the Hox genes led to a good deal of speculation on the function of the genes, their mode of action and the underlying reason(s) for the maintenance of the complex over a phylogenetic range including both the protostomes and deuterostomes. The period in the early 1980s to the mid 1990s when this speculation was rife has been referred to as the era of Homeomadness and was unfortunately characterized by what can be charitably called excessive behavior by some of the participants. Fortunately, as this book will attest, this period has mercifully ended and more measured analyses and conclusions now characterize the investigation of Hox gene function, genomic organization and evolution.

The book is divided into three major sections. The first four chapters cover aspects of the regulation of Hox gene expression (Chapters 2, 3 and 4) and the structure and function of the now justifiably well-known homeobox (Chapter 1). The second section offers insights and discussions of the sometimes contentious issues of the origin (Chapter 5) and evolution (Chapter 6) of the aforementioned Hox complexes. The two remaining chapters in this section (7 and 8) delve into the topic of the constraints on the conservation of the component loci of the complexes in animal phylogeny. In the third and last section the role of the resident loci in the specification of body plans and meristic identity (Chapters 9 and 10) of the arthropods is presented. The cases of Hox genes that have apparently gained novel functionality relative to their presumed ancestral roles in ontology are also noted. In addition to these discussions on segmental and tissue identity the role of the Hox genes in the specification of cellular identity in the nervous system is presented (Chapter 11). Finally the editor of the book shares his thoughts on the Hox genes and what constitutes inclusion as a member of this class of loci.

Wallace Arthur posits the following in his 2006 book, *Creatures of Accident: The Rise of the Animal Kingdom*, when noting the discovery of the homeobox:

> *In the 1980s two groups...made an important discovery. Actually this is putting it mildly: If I had to choose the most important biological discovery over the last half century, this would be it.*

Actually, the Hox genes, their complexes and the homeobox represent only the tip of the genetic conservation iceberg. The majority of developmentally important genes are conserved across all animal phyla. This latter discovery coming on the heels of the homeobox revelation has now altered the landscape of evolutionary thought and research. One must now be concerned with explaining how an essentially common set of molecular paradigms have been used to produce the vast array of extant and fossil animals. The chapters presented here are excellent examples of the change in landscape, thinking and research.

Prof. Thom C. Kaufman
Department of Biology
Indiana University at Bloomington
Bloomington, Indiana
USA

PREFACE

Are the Hox genes so important? Frankly, it is but a rhetorical question. In my opinion, the answer is YES, without any doubt. Still, I found it the right place to ask this question at the beginning of the present book. My purpose is here to argue, and try to convince those who may not agree, that the Hox genes are indeed so important that their study will not end with the 20th century.

In fact, I do think that the discovery of the homeobox[1,2] was one among the main scientific events in biology during the second half of the 20th century, together with the structure of the DNA molecule by Watson and Crick[3] and the discovery by Lwoff, Jacob and Monod[4] that the regulation of gene expression was itself genetically determined. I am old enough to remember the shocks that the discovery of the homeobox provoked a quarter of a century ago. Indeed, I can remember what our thoughts were at that time, and how much it changed our way of thinking, in molecular biology, developmental biology and evolutionary biology. In molecular biology, almost everybody at that time was convinced that each transcription factor had to be unique in both structure and sequence to accommodate its specific binding to DNA. Then came the Hox genes, and soon after the huge family of homeobox-containing transcription factors, contradicting this assumption. Similarly, although embryologists were familiar with the similarities in the development of mammals, and even between mammals and non-mammalian vertebrates, very few dared dream that studying a fly or a worm would bring deep insights to the knowledge of vertebrate development. Most biologists at that time were convinced that in order to make a fly different from a human being, despite clear homologies in the so-called 'house-keeping' genes, what would differ should be precisely developmental genes. Evolutionary biologists thought that evolution has produced new genes in order to achieve new body plans.[5] The same year when it was discovered in *Drosophila melanogaster*, the homeobox provided a means to retrieve homologous genes in other animals, including tetrapod vertebrates, mammalians and *Homo sapiens*.[6] In addition, it rapidly appeared that these homologous genes, now called 'Hox' genes, performed the same homeotic function in vertebrates as in flies.[7] This similarity brought into revival Geoffroy Saint-Hilaire's hypothesis of the "unity of design" throughout the whole animal kingdom,[8] an idea that seemed refuted after its famous dispute with Cuvier in 1830. Actually, it opened the door to a new discipline in biology, comparative developmental genetics, so-called 'evo-devo'.

The molecular problem is now coined 'the Hox paradox': How can a series of transcription factors, so similar in their homeodomain, the part of the molecule that binds to DNA, achieve the function of the Hox genes, which is precisely to give specificity—even often called 'identity'—to the domains in which they are expressed? This issue is not completely solved. Samir Merabet et al report the present-day answers and still-remaining questions in the first chapter of the present book.

The role of RNAs, including non-translated RNAs and microRNAs, in gene regulation has recently been brought into focus. Robert Maeda and François Karch examine the Bithorax-Complex (BX-C) of Hox genes in *Drosophila melanogaster* from this point of view and recall that Ed Lewis,[9] who can be seen as the founder of 'Hoxology', included as 'genes' in his genetic description of this complex, regulatory elements that are now known to correspond to such RNAs, the precise mechanisms of activity of which is still under research.

Another now fashionable stream in current research on gene expression is 'epigenetics'. In its modern meaning, epigenetics refers to the mechanisms by which gene expression is regulated without changes in the DNA sequence itself. Again, Hox genes provide a useful model. Indeed, expression of the Hox genes in Drosophila is maintained long after it has been initiated. Ed Lewis has studied the first known gene involved in Hox expression's maintenance.[9] These maintenance factors are, as well as the Hox genes themselves, preserved during metazoan evolution. In Chapter 3, Samantha Beck et al review the present knowledge about Hox genes' expression maintenance.

Another brilliant finding by Ed Lewis was what he called 'colinearity'. Indeed, he noticed that the genetic elements, genes and control elements, of the BX-C are arranged along the chromosome in the same order as their region of activity along the anterior to posterior (A-P) axis of the fly. This property is again conserved in most bilaterian metazoans. Hox genes cluster in complexes along the chromosomes in most bilaterians. François Spitz reviews in Chapter 4 the recent discoveries in the mammalian Hox complexes of regulatory regions that could account at least in part for the still mysterious colinearity property.

The second part of the book is devoted to the evolution of Hox genes and Hox complexes, in relation to animal evolution. Recent genomic data show that Hox-like genes are present in Cnidaria, thus predating the bilaterian radiation. Still, as argued by Bernd Schierwater and Kai Kamm in Chapter 5, the 'Hox system', defined as a complex of Hox genes involved in specifying the identity of the body regions along an axis, is specific to the Bilateria. The authors give an account of the debate on the origin of Hox genes and gene complexes. Then in Chapter 6, David Ferrier reviews the evolution of Hox genes and complexes, from their origin in cnidarians, to their diversification in bilaterians through duplications and losses. The theme of loss is exemplified by the Nematode case in Chapter 7 by Aziz Aboobaker and Mark Blaxter, and that of duplications by the case of the so-called 'posterior' Hox genes in Deuterostomes by Rob Lanfear in Chapter 8.

The third part of the book is devoted to the function of the Hox genes at the organism level. Wim Damen reviews in Chapter 9 the typical role of Hox genes as 'architect genes'[10] in the design and evolution of chelicerate body plans. In the 'Origin of Species' Darwin viewed the change of function of homologous organs during evolution as "so important".[11] Change of function was later theorized by Louis Cuénot as 'preadaptation'[12] and by Stephen J. Gould and Elizabeth Vrba as 'exaptation'.[12] In the Hox field, exaptation is exemplified in Chapter 10 by Urs Schmidt-Ott et al. by the evolution of the *Hox3/zen* gene in Insects. In Chapter 11, Heinrich Reichert and Bruno Bello review the function of the Hox genes in the design of the Drosophila brain. In the last Chapter, I discuss the biological function of the Hox genes in the Bilateria, and suggest that their morphological function, resulting in homeosis when the Hox genes are altered, is derived from a primitive neurogenic function that arose with the constitution of a central nervous system organized along the A-P axis at the origin of the Bilateria.

Prof. Emer. Jean S. Deutsch
Developmental Biology, Pierre and Marie Curie University
Paris, France

REFERENCES

1. McGinnis W, Levine MS, Hafen E et al. A conserved DNA sequence in homoeotic genes of the Drosophila Antennapedia and Bithorax Complexes. Nature 1984; 308:28-433.
2. Scott MP, Weiner AJ. Structural relationships among genes that control development: sequence homology between the Antennapedia, Ultrabithorax and fushi tarazu loci in Drosophila. Proc Natl Acad Sci USA 1984; 81:4115-4119.
3. Watson JD, Crick FHC. Molecular structure of nucleic acids. A structure of deoxyribose nucleic acid. Nature 1953; 171:737-738.
4. Jacob F, Monod J. Genetic regulatory mechanisms in the synthesis of proteins. J Mol Biol 1961; 3:318-356.
5. Gould SJ. The Structure of Evolutionary Theory. Cambridge: The Belknap Press of Harvard University Press, 2002.
6. McGinnis W, Garber RL, Wirz J et al. A homologous protein-coding sequence in Drosophila homeotic genes and its conservation in other metazoans. Cell 1984; 37:403-408.
7. Kessel M, Balling R, Grusss P. Variations of cervical vertebrae after expression of a Hox1.1 transgene in mice. Cell 1990; 61:301-308.
8. Geoffroy Saint-Hilaire E. Considérations générales sur la vertèbre. Mém Mus Hist Nat 1822; 9:89-114.
9. Lewis EB. A gene complex controlling segmentation in Drosophila. Nature 1978; 276:565-570.
10. McGinnis W, Kuziora M. The molecular architects of body design. Sci Am 1994; 270:58-66.
11. Cuénot L. Théorie de la préadaptation. Bologne: Scientia, 1914:59-73.
12. Gould SJ, Vrba ES. Exaptation - a missing term in the science of form. Paleobiol 1982; 8:4-15.

ABOUT THE EDITOR...

JEAN S. DEUTSCH, is a Professor Emeritus of Genetics and Animal Biology, Université Pierre et Marie Curie, Paris 6, Department (UMR 7622) "Biologie du Développement". Under the supervision of Prof. P.P. Slonimski, he participated to the birth of mitochondrial genetics of the yeast *Saccharomyces cerevisiae.* During the '80s, he moved to the Institut Jacques Monod in Paris to study developmental genetics of *Drosophila melanogaster*, focusing on the genetics of the hormonal control of metamorphosis. In 1993, he was the first in France, together with André Adoutte, to undertake evo-devo studies, choosing the cirripedes, which have been Darwin's favourite animals, as a model, because of their so peculiar body plan. In a second step, his team studied the developmental genetics of other arthropods, including scorpions and pycnogonids. He is author of a number of scientific publications in international journals, and of three textbooks in French on Drosophila and genetics.

PARTICIPANTS

Aziz Aboobaker
Institute of Genetics
The University of Nottingham
Queen's Medical Centre
Nottingham
UK

Samantha Beck
Molecular Epigenetics Group
Department of Zoology
Life Sciences Center
University of British Columbia
Vancouver, British Columbia
Canada

Bruno Bello
University of Basel
Basel
Switzerland

Mark Blaxter
Institute of Evolutionary Biology
University of Edinburgh
Edinburgh
UK

Hugh Brock
Molecular Epigenetics Group
Department of Zoology
Life Sciences Center
University of British Columbia
Vancouver, British Columbia
Canada

Wim G.M. Damen
Department of Genetics
Friedrich-Schiller-University Jena
Jena
Germany

Jean S. Deutsch
Developmental Biology
Pierre and Marie Curie University
Paris
France

Floria Faradji
Developmental Biology
Pierre and Marie Curie University
Paris
France

David E.K. Ferrier
The Scottish Oceans Institute
University of St Andrews
East Sands, St Andrews
UK

Yacine Graba
Institute of Developmental Biology
of Marseille Luminy
University of the Mediterranean
Marseille Cedex
France

Kai Kamm
ITZ, Ecology and Evolution
School of Veterinary Medicine
Hannover
Germany

François Karch
NCCR Frontiers in Genetics
and
Departement of Zoology
and Animal Biology
University of Geneva
Geneva
Switzerland

Thom C. Kaufman
Department of Biology
Indiana University at Bloomington
Bloomington, Indiana
USA

Robert Lanfear
School of Botany and Zoology
Center for Macroevolution
and Macroecology
Australian National University
Canberra
Australia

Steffen Lemke
Department of Organismal Biology
and Anatomy
University of Chicago
Chicago, Illinois
USA

Robert K. Maeda
NCCR Frontiers in Genetics
and
Departement of Zoology
and Animal Biology
University of Geneva
Geneva
Switrzerland

Samir Merabet
Institute of Developmental Biology
of Marseille Luminy
University of the Mediterranean
Marseille
France

Frédérique Peronnet
Developmental Biology
Pierre and Marie Curie University
Paris
France

Jacques Pradel
Institute of Developmental Biology
of Marseille Luminy
University of the Mediterranean
Marseille
France

Ab. Matteen Rafiqi
Department of Organismal Biology
and Anatomy
University of Chicago
Chicago, Illinois
USA

Heinrich Reichert
University of Basel
Basel
Switzerland

Nagraj Sambrani
Institute of Developmental Biology of Marseille Luminy
University of the Mediterranean
Marseille
France

Bernd Schierwater
ITZ, Ecology and Evolution
School of Veterinary Medicine
Hannover
Germany

Urs Schmidt-Ott
Department of Organismal Biology and Anatomy
University of Chicago
Chicago, Illinois
USA

François Spitz
Developmental Biology Unit
European Molecular Biology Laboratory
Heidelberg
Germany

CONTENTS

SECTION I. MECHANISMS OF ACTIVITY

ACKNOWLEDGEMENTS

Each chapter of the present book has been reviewed by two referees, one being a contributor of another chapter, and the second being an external expert. The editor wishes here to thank these distinguished scientists for their contribution to the book: Giacommo Cavalli, Jacqueline Deschamps, Claude Desplan, Brigitte Galliot, Volker Hartenstein, Thom Kaufman, Shige Kuraku, Michael Manuel, Kristen Panfilio, Eric Queinnec, Max Telford.

Jean S. Deutsch

Section I
Mechanisms of Activity

Chapter 1

Regulation of Hox Activity: Insights from Protein Motifs

Samir Merabet,* Nagraj Sambrani, Jacques Pradel and Yacine Graba*

Abstract

Deciphering the molecular bases of animal body plan construction is a central question in developmental and evolutionary biology. Genome analyses of a number of metazoans indicate that widely conserved regulatory molecules underlie the amazing diversity of animal body plans, suggesting that these molecules are reiteratively used for multiple purposes. Hox proteins constitute a good example of such molecules and provide the framework to address the mechanisms underlying transcriptional specificity and diversity in development and evolution. Here we examine the current knowledge of the molecular bases of Hox-mediated transcriptional control, focusing on how this control is encoded within protein sequences and structures. The survey suggests that the homeodomain is part of an extended multifunctional unit coordinating DNA binding and activity regulation and highlights the need for further advances in our understanding of Hox protein activity.

Introduction

Hox genes, differentially expressed along the anterior-posterior (A-P) body axis, play fundamental roles in organising animal body plans.[1-3] Hox genes have been evolutionarily conserved among bilaterians at the organisational, structural and functional levels: usually clustered in complexes, they encode transcription factors and provide during development A-P axial positional information.[1,4] It is recognised that variation in Hox genes number, expression pattern and Hox protein activities have played a major role in the evolution of metazoan body plan.[2,5] The Hox complement is organized in paralog groups (seven to nine in the common ancestor of bilaterians and fourteen in vertebrates) each one corresponding to highly related genes that share an intimate evolutionarily history, display strong sequence similarity, occupy equivalent positions within the genomic clusters and exhibit similar expression patterns and functions. Hox genes have also been associated to a number of human diseases.[6-9] Thus, the importance of Hox protein function in development, evolution and physio-pathological processes is well established.

Better understanding Hox protein mode of action requires advances into two main directions. The first direction should decipher how Hox proteins control organogenesis. Organs correspond to well-organised three dimensional arrays of different cell-subtypes that ultimately fulfil distinct physiological functions. In controlling segment-specific organogenesis, Hox proteins play major roles by coordinating the organisation and differentiation of pluricellular structures.[10] As Hox genes encode transcription factors, this is believed to rely on the selective regulation of downstream gene networks that control diverse cellular processes underlying organogenesis.[11-14]

*Corresponding Authors: Samir Merabet and Yacine Graba—Institut de Biologie du Dévelopement de Marseille Luminy, IBDML, UMR 6216, CNRS, Université de la méditerranée, Parc Scientifique de Luminy, Case 907, 13288, Marseille Cedex 09, France. Emails: merabet@ibdml.univ-mrs.fr, graba@ibdml.univ-mrs.fr

Hox Genes: Studies from the 20th to the 21st Century, edited by Jean S. Deutsch.

The second main direction should unravel general principles and mechanisms underlying transcriptional regulation by Hox proteins. Hox proteins all share a highly conserved helix turn helix DNA-binding motif, the homeodomain (HD) and consequently recognise a highly frequent and similar if not identical consensus DNA sequence (TAAT).[15] Hox proteins also share an additional signature, the hexapeptide (HX), a six amino acid sequence located N-terminal to the HD. This motif mediates the interaction with the Exd/Pbx class of cofactors (collectively referred to as PBC), which was shown to improve Hox DNA-binding affinity and selectivity.[16]

Approaches to identify protein motifs involved in the control of Hox transcriptional activity, with the aim of deciphering the molecular cues that distinguish the activity of one versus another Hox protein, were long focused on the use of chimeric Hox (for examples see refs. 17-22). Surprisingly, functional dissections of Hox proteins were rarely approached by direct motif mutations, which may in part be explained by the general belief that impairing a functional motif by point mutations should result in a generally inactive protein. Recent work however proved this belief to be wrong, showing that even mutation of highly conserved motifs like the HX does not result in globally inactive proteins, pioneering the road towards more direct and may be easily interpretable approaches for the identification of Hox functional protein motifs.

Here we summarize our current knowledge on the molecular bases of Hox-mediated transcriptional control, focusing on how this regulation is encoded within Hox protein sequences, including obligatory or widely shared domains (HD and HX) as well as additional less conserved motifs.

The Homeodomain

Phylogeny of Hox Genes and General Features of Hox Homeodomains

HD-containing proteins are regrouped in several classes. Among them, the ANTP superclass is composed of the closely related NK and Hox genes. While NK genes, as well as representatives of five other HD classes (LIM, TALE, PRD, POU and SIX) exist in sponges, Hox genes are first present in the common ancestor of Cnidaria and Bilateria (Fig. 1). This led to the proposition that NK genes could be at the origin of Hox genes.[23] Alternatively, Hox genes could have originated from a proto-Hox gene present in the ancestor of all metazoans, with a loss in sponges like *Amphimedon*.[24]

Hox genes are generally assumed to provide axial positional identity. Such a function is however not clearly established in Cnidarians,[25-27] which are organized along an oral-aboral axis whose homology with the bilaterian A-P axis remains unclear. The function of Hox genes in providing axial identity most likely was acquired in bilaterians, which correlates with the acquisition of a segmental mode of embryonic development and with the huge diversification of bilaterian body plans.[28]

HDs are all composed of three alpha helices that are preceded by a flexible N-terminal arm (Fig. 2). Helices 2 and 3 form a helix-turn-helix (HTH) motif, an ancestral DNA-binding motif also found in prokaryotic proteins that are not transcription factors, like gamma transposases or Hin recombinases.[15,29-30] The three helices form a compact structure where helices 1 and 2 are anti-parallel and pack against helix 3 in a perpendicular orientation. Helix 3 fits into the major groove of DNA with its hydrophilic face in a roughly parallel orientation. Residues from the loop between helices 1 and 2 and at the beginning of helix 2 engage contacts with the DNA backbone. However, most of the protein-DNA contacts and more particularly base-specific contacts, occur with helix 3, also called "recognition helix" and with the N-terminal arm that fits into the minor groove.[15] Hydrophobic core amino acids that define the tight three-dimensional conformation of the HD and basic amino acids that contact DNA are under structural constraints and are highly conserved across all classes of HD-containing transcription factors. Consequently, divergent HDs adopt a remarkable similar overall arrangement when bound to DNA (Fig. 2). These structural features emphasize a generic mode of DNA binding, where Hox and non-Hox HDs share the same main characteristics.

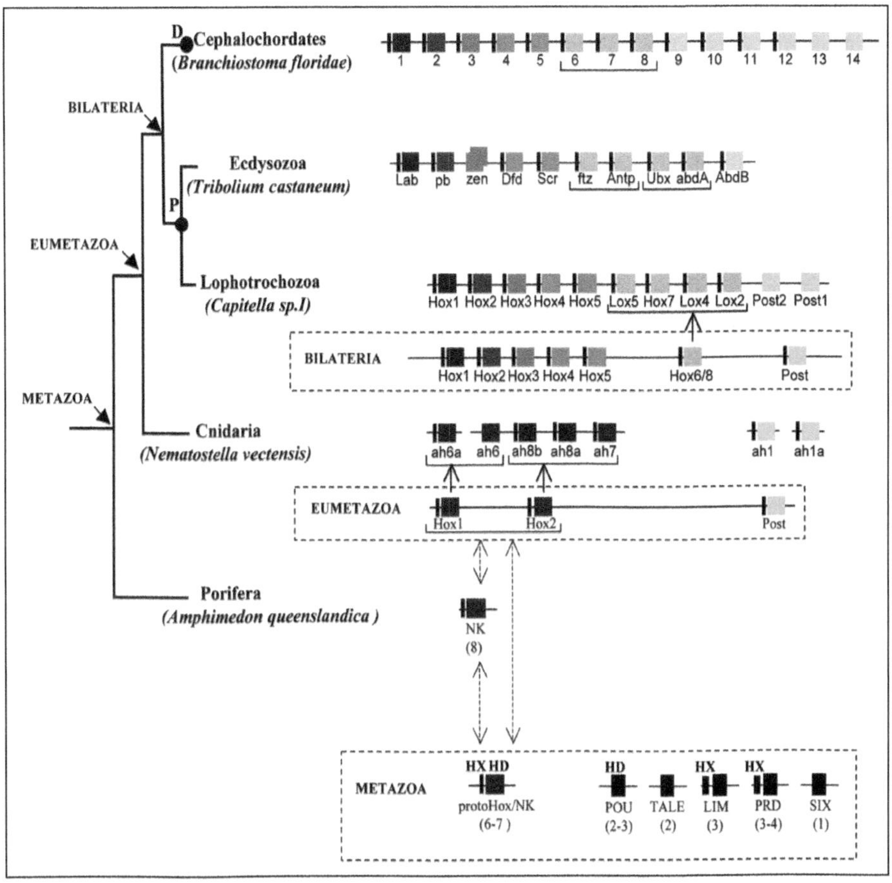

Figure 1. Phylogenetic distribution of the HX motif within Hox and HD-containing genes. The main branches of the animal evolutionary tree are indicated together with the Hox complements. Hox genes are symbolized by grey and black boxes that respectively represent the HD and HX. In other HD classes, the HD and HX, when present, are symbolized by a black box. Hox complexes present in hypothetical ancestors of bilaterians, eumetazoans and metazoans are shown in dotted boxes. Number of representatives for each HD class is indicated in brackets.[23] Arrows point possible duplication events; for example, Hox6/8 from the ancestral Bilateria complex duplicated independently in Protostomia (P) and Deuterostomia (D) to generate central paralog Hox genes. Hypothetical events giving rise to Hox genes in the Eumetazoa ancestor (see text) are indicated by dotted arrows. The main other HD classes are present in the Metazoa ancestor.

Specificity of Hox Homeodomain DNA Binding

Since many HDs show similar nucleotide binding sequence preferences,[31] they have long been considered as poorly stringent DNA binding motifs, raising the issue of specificity in the DNA recognition modes. Two recent studies have undertaken systematic analyses for the DNA binding properties of most fly and mouse HDs.[32,33] These studies revealed that HDs discriminate between subtle variations in cognate target site, which led to a classification in several specificity groups.

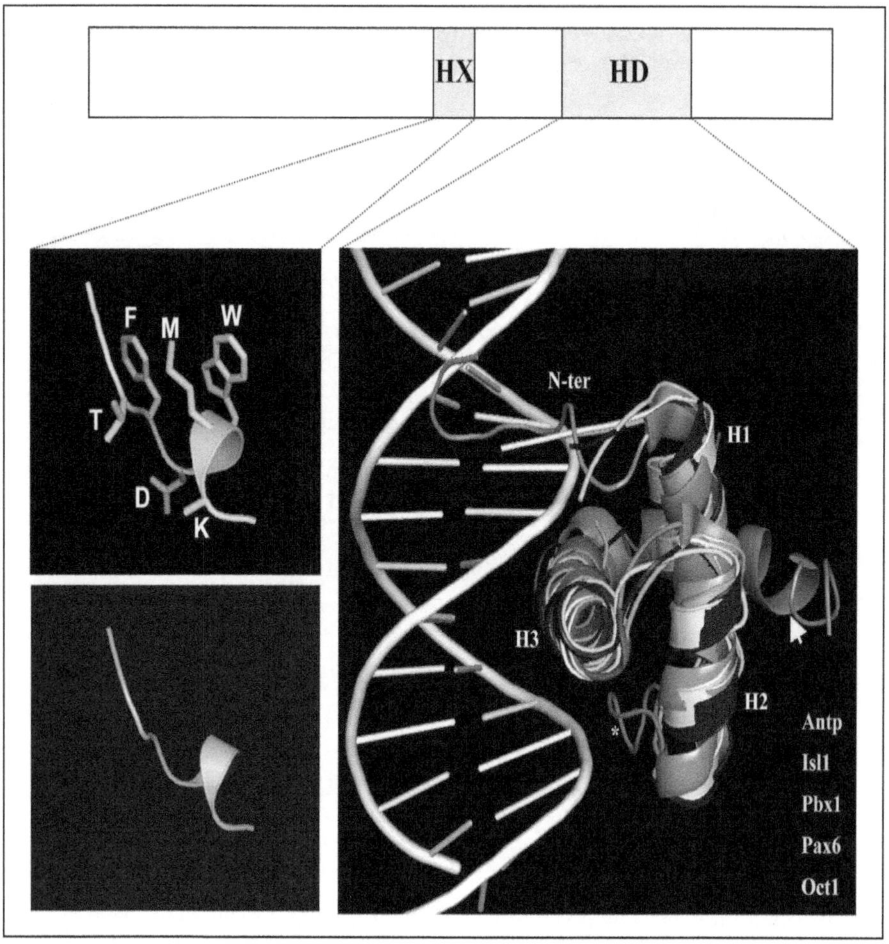

Figure 2. Conserved HD DNA binding modes. Upper panel schematizes the general architecture of a Hox protein, with a conserved HX and HD. Left panels: Structure of the HX motif illustrated here by the HoxB1 HX. The motif forms a helix. Schematic representations realized with the PyMol Viewer software, displaying (upper panels) or not (lower panels) amino acids side chains. Right panel: Superimposition of structures of DNA-bound HDs from proteins of five different classes: Antennapedia (Antp, Antp class), Islet1 (Isl1, LIM class), Pbx1 (TALE class), Pax6 (paired class) and Oct1 (POU class). The schematic representation was realized by assembling 3D-structures of the HDs (http://www.rcsb.org/pdb/home/home.do) with the PyMol Viewer software. N-terminal arms inserted within the DNA minor groove and the three helices (H1-3) of the HDs are indicated. Pbx TALE and a structured C-terminal extension are highlighted by a star and white arrow respectively.

Hox proteins fall into two distinct specificity groups: one that encompasses paralog groups 1 to 8 and the other that corresponds to the posterior paralog groups. Proteins of the first group preferentially recognize the canonical TAAT core sequence, while the posterior ones bind to an unusual TTAT core site.[34] The consensus sequence designed for all Hox proteins is $T_1N_2A_3T_4N_5N_6$, where N is variable. Positions 3 to 6 lie in the DNA major groove and are contacted by residues 47, 50, 51 and 54 of the recognition helix.[35] These residues are identical in all Hox proteins (with a conservative change at position 47 in paralog group 2), explaining their common capacity to recognize "AT rich" sequences. Positions 1 and 2 are localized in the minor groove and are contacted by residues 3 and 5 of the N-terminal arm of the HD, which are less conserved than DNA contacting residue of the recognition helix. Support for the general importance of the N-terminal arm of the HD came from in vivo analyses demonstrating its contribution in providing functional specificity to the *Drosophila* Hox proteins Sex combs reduced (Scr),[20] Antp,[21] Ultrabithorax (Ubx)[19,36] and AbdominalA (AbdA).[22]

DNA binding is also influenced by base nonspecific contacts and the topology of the target sequence. This is well documented for the human HoxA9 HD, which displays a large number of interactions with the DNA backbone and may explain the overall increased DNA affinities of posterior Hox proteins.[37] The importance of DNA topology has been recently highlighted by the crystallographic resolution of a Scr/DNA complex on a physiological target sequence.[38] In that structure, the target sequence presents a particularly narrow minor groove ideally suited for the insertion of the N-terminal arm of the Scr HD.

In summary, the DNA-binding mode of Hox proteins relies on multiple molecular events, including both specific and nonspecific interactions with DNA bases. These two recognition processes neither involve a specific domain in the HD, nor act in autonomous ways. For example, interactions between the recognition helix and the DNA major groove are influenced at a distance by the positioning of the N-terminal arm within the DNA minor groove.[39] These observations compromise the existence of a simple recognition code whereby the identity of Hox DNA binding target sites could be inferred by the primary sequence of the HD.

Activity Regulation of Hox Homeodomains

Transcriptional regulation by sequence specific transcription factors relies on two conceptually distinguishable steps: DNA binding, which defines target gene selection; and activity regulation, which defines the level of target gene activation or repression. The importance of the second step has been well demonstrated by in vivo studies of *Drosophila* Hox proteins fused to the potent activation domain VP16, where a modification in activity regulation was found sufficient to change Hox functional identity.[40,41]

The HD can modulate Hox activity regulation by a variety of molecular mechanisms. In the case of the *Drosophila* Deformed (Dfd), Ubx[41] and mouse HoxA13 proteins,[42] the HD indirectly controls the regulatory potential by masking transcriptional activation domains. Hox HDs can also indirectly antagonize activating functions through heterodimeric interactions. This is exemplified by the human HoxD8 protein, which counteracts the activating potential of HoxA9.[43] This interaction does not affect the DNA-binding properties of HoxA9 and requires both the N-terminal arm and helix 1 of HoxD8 HD. Finally, Hox HDs can also behave as potent repressors, through distinct regions: repressive functions of mouse HoxA7 are provided by the recognition helix,[44] while repression by HoxA11 involves the N-terminal arm and helices I and II of the HD.[45]

Mapping on crystal structures reveals that residues involved in activity regulation are often oriented away from the DNA double helix and in position available for protein-protein interactions. So far however, no factor interacting specifically with these residues has been unambiguously identified. One candidate interacting protein that may control Hox activity regulation is the histone acetyltransferase (HAT) CBP, a well known transcriptional activator. In Hox-CBP dimers,[46-50] the two factors are mutually antagonist: CBP impairs Hox DNA-binding activities and Hox proteins inhibit CBP HAT activity. This inhibition could account for Hox-mediated repression. Additional cofactors recruited by the HD of Hox proteins have been described that either enhance[42,51-54] or reduce[55-57] Hox transcriptional activity. The increasing number of interactions established by Hox HDs with cofactors suggests that protein-protein interactions are essential for defining the final Hox transcriptional output.[55]

Homeodomain-Mediated Transport

Hox proteins have the capacity to shuttle between the cytoplasm and nucleus, a process that relies on HD sequences. Residues important for establishing DNA contacts also define nuclear localization signals.[58] Surprisingly, the HD also carries the capacity to cross the plasma membrane.[59] This unanticipated function was first established for Antp, as well as for vertebrate HoxA5 HDs. Both were shown to be taken up from the medium by cultured neurons, fibroblasts and myoblasts.[60,61] The cellular uptake of the HD relies on the third helix and happens through an unconventional, energy free, vesicle and receptor independent internalization process.[62] While the physiological importance of this mechanism was recently established for the Otx2 homeoprotein in the control of postnatal neuronal plasticity in the visual cortex,[63] evidence for a membrane crossing function of the Hox HDs in developmental contexts is still lacking.

The Hexapeptide Motif

Cofactor Mediated Control of Hox Target Gene Specificity

Apart from the HD, the only protein motif widely conserved in Hox proteins is a short peptide, termed hexapeptide (HX), pentapetide or more recently PID (for PBC Interacting Domain).[64] The HX lies upstream of the HD, contains a core sequence centered around an invariant tryptophan and folds into a short helicoidal structure[65] (Fig. 2). HX found in posterior paralog groups usually only retain the central tryptophan, known to play a key role in establishing contacts with PBC proteins.[37,65]

The phylogenetic distribution of the HX motif within HD containing proteins shows that the motif most likely appeared in the common ancestor of metazoans in the Antp superclass, including Hox and NK class HDs (Fig. 1). A more divergent, possibly HX related motif, was found in the LIM and PRD HD-containing proteins, as well as in the bHLH transcription factor MyoD.[64]

The Hox HX motif was early shown to mediate interaction with PBC proteins. PBC are part of the TALE (Three Amino Acid Loop Extension: Fig. 2) HD class, which is characterized by a three-residue insertion in between helices 1 and 2.[66] Representatives of PBC proteins are highly conserved[67] and includes *Drosophila* Exd and vertebrate Pbx proteins. Exd was initially described from homeotic phenotypes associated to its mutation[68] and Pbx1 from a translocation resulting in a fusion protein responsible for preB cell acute lymphoblastic leukemia in human.[69] Initial support for a role of the HX in PBC recruitment came from in vitro protein interaction assays,[70-73] which established that both Hox and PBC partners cooperatively bind DNA. Crystallographic studies[37,38,65,74] remarkably confirmed the structural model raised from initial biochemical analyses.[75] The HD of Hox and PBC proteins face each other in a head to tail orientation, with the PBC HD contacting the 5' nucleotides of the binding site in the minor groove, while the Hox HD binds the 3' half of the site in the major groove. Protein-protein interactions are mainly, although not exclusively, mediated by the HX motif on the Hox side and by a hydrophobic pocket formed in part by the TALE on the PBC side.

Formation of Hox-PBC complexes both improves DNA-binding affinities, especially for anterior and central groups of Hox proteins[76] and extends the size of cognate DNA sequences, therefore allowing higher specificity in target recognition.[16,75,77,78] This generic mode of Hox-PBC interaction provided the basis for a model underlying Hox DNA binding specificity, whereby subtle differences in central positions of the Hox-PBC target sequence define the identity of the Hox protein engaged in the Hox-PBC complex.[79-81] Selective DNA-binding may be brought in by HX surrounding sequences, which were shown to influence the DNA recognition mode by the HD N-terminal arm.[38] Of note, sequences surrounding the HX are well conserved and sufficient to classify Hox proteins in appropriate paralog groups.[38,82]

Additional complexities certainly underlie Hox DNA binding specificity, as illustrated by the fact that the *Drosophila* Labial (Lab)/Exd complex recognizes very divergent target sequences[83] and by the more recent finding that the topology of an AbdA-Exd target site may diverge from the consensual Hox-PBC sequence.[84] Finally, a peculiar function for the HX was found in Lab, where it acts as an inhibitory DNA binding motif by preventing the HD to interact with DNA. In this case, the recruitment of Exd induces a conformational change that neutralizes the inhibitory role of the HX and allows cooperative binding of the Lab/Exd complex to DNA.[85]

PBC-Independent Functions for the Hexapeptide Motif

So far, in vivo analyses of HX mutation have been performed for seven Hox proteins, including the Fushi tarazu (Ftz) protein, a central arthropod paralog at the border of homeotic function.[86,87] Some of the associated phenotypes, for mouse HoxA1 in hindbrain and skeletal specification, or for *Drosophila* Ubx in segment identity specification[88] and Ftz,[86,87] are consistent with a PBC recruiting function. Other phenotypes are however difficult to reconcile[89] or even incompatible with a function of the HX in mediating PBC recruitment.[88,90] This is well demonstrated by mutation of the HX in Ubx, which neither impairs Exd-dependent *Distalless* (*Dll*) repression, nor alleviates Exd recruitment in vitro.[88,90,91] The HX thus fulfills additional functions distinct from recruiting Exd. This conclusion was reinforced by the observation that HX mutation in AbdA converts the protein from a repressive to an active form.[90] This qualitative change in AbdA activity regulation is Exd-independent, suggesting that the activity regulation function of the HX motif likely relies on interactions with other protein partners. Recently, such a cofactor for Antp has been identified, whose nature (a TATA-binding protein associated factor) establishes an intimate link between Hox protein function and the general transcriptional machinery.[92]

Additional Hox Functional Motifs

Alternate Motifs for PBC Recruitment?

Hox-PBC interactions can occur in absence of the HX,[88,90] raising the question of what motifs could be alternatively used. A first indication came from the observation that residues located within and C-terminal to the HD of Ubx contribute to PBC interactions.[93,94] More recent evidence originates from studies of a motif lying just C-terminally to the HD. This motif has been called "UbdA" and its mutation in *Drosophila* Ubx abolishes both Exd recruitment in vitro and Exd-dependent repression in vivo.[91] The UbdA motif, which is absent in deuterostomes, constitutes both a clade and paralog specific motif.[29] These findings establish that HX-alternative modes of PBC recruitment do exist, introducing flexibility in Hox-PBC-DNA contacts. Resolving the structure of UbdA-mediated Exd interaction should provide novel cues to explain how transcriptional diversity is generated through qualitatively distinct interaction modes involving the same protein partners.

To what extent such additional, non-generic and paralog-specific PBC interaction modes exist in other Hox proteins still remains an open question. It was reported that a HX-deficient Lab protein retains Exd interaction potential and in vivo Exd-dependent activity.[95] As Lab does not contain an UbdA-like motif, Exd recruitment should occur through a yet unknown motif, supporting further the notion of flexibility in Hox-PBC contacts.

The Linker Region: A Variable but Crucial Hox Protein Domain

Most of the efforts to understand the molecular cues underlying Hox protein function have long focused on the HD and HX, whereas the linker region (LR), the sequence connecting the HX and HD , was considered as a variable, unstructured and neutral region.[96]

LR size does not vary randomly. It is roughly comparable within each paralog group, with the more posterior Hox proteins containing gradually shorter LRs.[64] Interestingly, short LR present ordered structures that do not impose constraints weakening DNA binding affinity.[37] In contrast, more anterior Hox proteins display longer and structurally disordered LR,[65] which imposes a stronger requirement for cofactor recruitment to achieve efficient DNA binding.

LR size also varies between distinct spliced products of the same Hox gene. This is the case for several *Drosophila* Hox genes and is best exemplified for *Ubx*, where alternative splicing of three micro-exons within the LR sequence generates six distinct proteins. These isoforms differentially interact with the Exd cofactor,[97] are evolutionarily conserved in different *Drosophila* species[98] and their production is developmentally regulated,[99] suggesting that they may convey distinct biological functions.[100]

Evidence for the functional importance of the LR in controlling Hox protein functions initially came from the observation that phosphorylation sites within the LR of the mouse HoxB7 were crucial for its inhibitory activity on granulocytes differentiation.[101] Further evidence was obtained from *Drosophila* developmental studies. First, different Ubx isoforms repress to different extents the *Dll* target gene and it has been suggested that the LR was important for the repressive potential of Ubx.[93] Second, an evolutionary conserved motif lying within the LR of AbdA (the PFER motif) was also shown to control the repressive potential of AbdA independently of its DNA-binding activities.[90] Finally, the study of the regulation of the *forkhead* target gene by Scr showed that critical residues for proper function lie within the LR, which adopts an ordered structure and directly contribute to Scr DNA binding specificity.[38]

Transcriptional Activation and Repression Domains

Hox proteins act both as transcriptional activators and repressors. Accordingly, activation and repression domains have been identified in Hox proteins.[51,52,102-105] One such protein domain containing a core SSYF peptide is present in several paralog groups and highly conserved in both vertebrates and invertebrates. Deleting this N-terminal domain in Ubx, Scr[104] or HoxA5[102] leads to a global loss of transcriptional activation.[104]

Regulation of the Hox transcriptional repression potential was also proposed to play a major role in the evolution of pan-arthropod body plan organization. One salient regulated morphological diversification concerns the number of appendices: the onychophoran *Akanthokara kaputensis* and the crustacean *Artemia franciscana* develop limbs all along the body, while the insect *Drosophila melanogaster* does not. It was proposed that the underlying mechanisms, although different in *Akanthokara kaputensis* and *Artemia franciscana*, rely on Hox activity regulation: coexistence of Ubx and limb formation is explained in *Akanthokara kaputensis* by the lack of a repression domain (a QAQA core motif followed by an alanine-rich sequence[106,107] and in *Artemia Franciscana* by the existence of an intermediate regulatory module (composed of a series of serine and threonine residues) that inhibits limb repressive activities.[106,107]

Other activation or repression domains exist and were mostly identified through cell culture transcriptional assays. These domains are ill-defined, lacking canonical signatures for activation or repression and usually display low sequence evolutionary conservation. Some of them act depending on the cellular or promoter context, either as activation or repression domains[103,105] and are regulated by posttranslational modifications.[108,109]

The importance of such regulatory modules in controlling Hox functions is also revealed by the existence of several mutations affecting Hox coding sequences in mice or humans. This is best illustrated by limbs and genitourinary tract malformations (collectively referred as hand-foot-genital syndrome, HFGS), which result from an abnormal expansion of an alanine-coding repeat localized in the N terminal parts of HoxA13 and HoxD13 proteins.[110,111]

Conclusion

While our survey discusses how functional modules within Hox proteins contribute to their function, we feel it is important to emphasize that the view provided is obviously limited by the few modules identified so far, as well as by the restricted windows within which they have been analysed. Further work is required to assess whether Hox function requires a small or large set of intrinsic regulatory modules and to understand how these modules are used to generate the specificity and diversity underlying Hox protein function in development and evolution.

Most of the regulatory modules identified so far are clustered around the HD (Fig. 3A). The HD and surrounding sequences thus constitute an integrated multifunctional regulatory unit that controls both DNA-binding and activity regulation of Hox proteins. This dual role

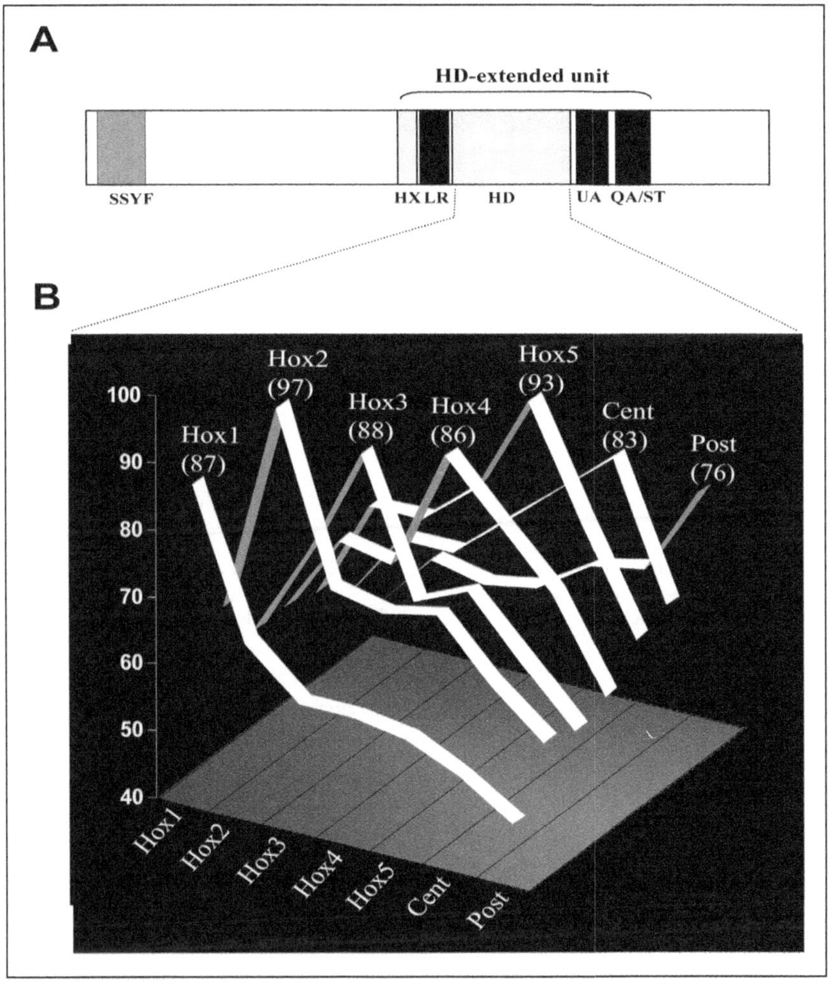

Figure 3. A multifunctional unit anchored on a highly conserved HD. A) Functionally important motifs are clustered around the HD. The scheme integrates motifs found to contribute to arthropod Ubx activity, as discussed in the text. Except the SSYF activation domain, all other motifs are closely associated to HD sequences. HX, LR, HD, UA, QA, ST respectively stands for hexapeptide, linker region, homeodomain, UbdA, QAQA motif and serine/threonine rich domain. Light grey, grey and black boxes respectively highlight motifs shared by all Hox proteins, by some paralog groups only, or motifs specific to central paralogs. B) HD sequence conservation between Hox paralog groups. Percentage of identity progressively decreases as distant paralog groups of the *Hox* cluster are compared. Central paralog group corresponds to *Hox 6-8* genes, while posterior paralog group corresponds to *Hox 9-10* genes (sequences from paralog groups 11-13, which are not found in all the major branches of bilaterians, were voluntary not kept). Representatives sequences of the three main bilateria branches were taken: Deuterostomia/NP_034579, NP_034581, NP_034582, NP_032291, NP_034583, NP_034584, NP_034585, NP_034591, NP_034586, NP_032289; Ecdysozoa: AAD46166, AAD46167, AAD46168, AAB61441, AAD46170, AAD46174, AAD46171, AAD46173, AAD46172, AAD46175, AAD46176; Lophotrochozoa/NP_001107762, NP_001107807, NP_001036813, AAK16423, AAK16422, NP_001034505, NP_001034497, NP_001034518, NP_001034519.

of the HD extended unit may allow coordinate regulation/evolution of distinct regulatory mechanisms, which together define Hox protein function. Within this extended unit, highest conservations apply to the HD itself, with identity score within paralog group ranging from 97% to 76% (Fig. 3B). This high intra-paralog conservation can explain the striking functional equivalence of paralogous genes sharing less than 50% identity in protein coding sequences[112] and suggests that acquisition of specific features within paralog groups has largely contributed to define the regulatory potential of Hox proteins. Remarkably, comparing the paralog consensus two by two reveals that sequence conservation between adjacent groups is high and progressively decreases between more distant paralog groups (Fig. 3B). Inter-paralog conservation thus correlates with genomic colinearity of Hox complexes, possibly illustrating the history of Hox complex genesis.

Acknowledgements

Research in the author's laboratory is supported by the CNRS, Université de la Méditerranée and grants from CEFIPRA, ANR and ARC. We apologize to any investigator whose work owing to space limitation was not covered thoroughly or directly cited.

References

1. McGinnis W, Krumlauf R. Homeobox genes and axial patterning. Cell 1992; 68:283-302.
2. Gellon G, McGinnis W. Shaping animal body plans in development and evolution by modulation of Hox expression patterns. Bioessays 1998; 20:116-25.
3. Deutsch JS, Mouchel-Vielh E. Hox genes and the crustacean body plan. Bioessays 2003; 25:878-87.
4. Duboule D. The rise and fall of Hox gene clusters. Development 2007; 134:2549-60.
5. Warren R, Carroll S. Homeotic genes and diversification of the insect body plan. Curr Opin Genet Dev 1995; 5:459-465.
6. Argiropoulos B, Humphries RK. Hox genes in hematopoiesis and leukemogenesis. Oncogene 2007; 26:6766-76.
7. Eklund EA. The role of HOX genes in malignant myeloid disease. Curr Opin Hematol 2007; 14:85-9.
8. Cillo C, Cantile M, Faiella A et al. Homeobox genes in normal and malignant cells. J Cell Physiol 2001; 188:161-9.
9. Abate-Shen C. Deregulated homeobox gene expression in cancer: cause or consequence? Nat Rev Cancer 2002; 2:777-85.
10. Hombria JC, Lovegrove B. Beyond homeosis—HOX function in morphogenesis and organogenesis. Differentiation 2003; 71:461-76.
11. Pearson JC, Lemons D, McGinnis W. Modulating Hox gene functions during animal body patterning. Nat Rev Genet 2005; 6:893-904.
12. Hueber SD, Lohmann I. Shaping segments: Hox gene function in the genomic age. Bioessays 2008; 30:965-79.
13. Hueber SD, Bezdan D, Henz SR et al. Comparative analysis of Hox downstream genes in Drosophila. Development 2007; 134:381-92.
14. Graba Y, Aragnol D, Pradel J. Drosophila Hox complex downstream targets and the function of homeotic genes. BioEssays 1997; 19:379-388.
15. Gehring WJ, Qian YQ, Billeter M et al. Homeodomain-DNA recognition. Cell 1994; 78:211-223.
16. Mann RS, Chan S-K. Extra specificity from extradenticle: The partnership between HOX and PBX/EXD homeodomain proteins. Trends Genet 1996; 12:258-262.
17. Mann RS, Hogness DS. Functional dissection of Ultrabithorax proteins in D. melanogaster. Cell 1990; 60:597-610.
18. Chan S-K, Mann RS. The segment identity functions of Ultrabithorax are contained within its homeo domain and carboxy-terminal sequences. Genes Dev 1993; 7:796-811.
19. Lin L, McGinnis W. Mapping functional specificity in the Dfd and Ubx homeo domains. Genes Dev 1992; 6:1071-1081.
20. Zeng W, Andrew DJ, Mathies LD et al. Ectopic expression and function of the Antp and Scr homeotic genes: The N terminus of the homeodomain is critical to functional specificity. Development 1993; 118:339-352.
21. Furukubo-Tokunaga K, Flister S, Gehring WJ. Functional specificity of the Antennapedia homeodomain. Proc Natl Acad Sci Usa 1993; 90:6360-6364.
22. Chauvet S, Merabet S, Bilder D et al. Distinct Hox protein sequences determine specificity in different tissues. Proc Natl Acad Sci USA 2000; 97:4064-9.

23. Larroux C, Fahey B, Degnan SM, et al. The NK homeobox gene cluster predates the origin of Hox genes. Curr Biol 2007; 17:706-10.
24. Peterson KJ, Sperling EA. Poriferan ANTP genes: primitively simple or secondarily reduced? Evol Dev 2007; 9:405-8.
25. Jakob W, Schierwater B. Changing hydrozoan bauplans by silencing Hox-like genes. PLoS ONE 2007; 2:e694.
26. Kamm K, Schierwater B, Jakob W et al. Axial patterning and diversification in the cnidaria predate the Hox system. Curr Biol 2006; 16:920-6.
27. Chiori R, Jager M, Denker E et al. Are Hox genes ancestrally involved in axial patterning? Evidence from the hydrozoan Clytia hemisphaerica (Cnidaria). PLoS ONE 2009; 4:e4231.
28. Balavoine G, de Rosa R, Adoutte A. Hox clusters and bilaterian phylogeny. Mol Phylogenet Evol 2002; 24:366-73.
29. Feng JA, Johnson RC, Dickerson RE. Hin recombinase bound to DNA: the origin of specificity in major and minor groove interactions. Science 1994; 263:348-55.
30. Deutsch J, Lopez P. Are transposition events at the origin of the bilaterian Hox complexes? Evolving pathways, Cambridge University Press, 2008;13.
31. Gehring WJ, Affolter M, Bürglin T. Homeodomain proteins. Annu Rev Biochem 1994; 63:487-526.
32. Berger MF, Badis G, Gehrke AR et al. Variation in homeodomain DNA binding revealed by high-resolution analysis of sequence preferences. Cell 2008; 133:1266-76.
33. Noyes MB, Christensen RG, Wakabayashi A et al. Analysis of homeodomain specificities allows the family-wide prediction of preferred recognition sites. Cell 2008; 133:1277-89.
34. Ekker SC et al. The degree of variation in DNA sequence recognition among four Drosophila homeotic proteins. EMBO J 1994; 13:3551-3560.
35. Mann RS. The specificity of homeotic gene function. BioEssays 1995; 17:855-863.
36. Chan SK, Mann RS. The segment identity functions of Ultrabithorax are contained within its homeo domain and carboxy-terminal sequences. Genes Dev 1993; 7:796-811.
37. LaRonde-LeBlanc NA, Wolberger C. Structure of HoxA9 and Pbx1 bound to DNA: Hox hexapeptide and DNA recognition anterior to posterior. Genes Dev 2003; 17:2060-72.
38. Joshi R et al. Functional specificity of a Hox protein mediated by the recognition of minor groove structure. Cell 2007; 131:530-43.
39. Frazee RW, Taylor JA, Tullius TD. Interchange of DNA-binding modes in the Deformed and Ultrabithorax homeodomains: a structural role for the N-terminal arm. J Mol Biol 2002; 323:665-83.
40. Li X, McGinnis W. Activity regulation of Hox proteins, a mechanism for altering functional specificity in development and evolution. Proc Natl Acad Sci USA 1999; 96:6802-7.
41. Li X, Murre C, McGinnis W. Activity regulation of a Hox protein and a role for the homeodomain in inhibiting transcriptional activation. Embo J 1999; 18:198-211.
42. Suzuki M, Ueno N, Kuroiwa A. Hox proteins functionally cooperate with the GC box-binding protein system through distinct domains. J Biol Chem 2003; 278:30148-56.
43. Zappavigna V, Sartori D, Mavilio F. Specificity of HOX protein function depends on DNA-protein and protein-protein interactions, both mediated by the homeo domain. Genes Dev 1994; 8:732-744.
44. Schnabel CA, Abate-Shen C. Repression by HoxA7 is mediated by the homeodomain and the modulatory action of its N-terminal-arm residues. Mol Cell Biol 1996; 16:2678-2688.
45. Roth JJ, Breitenbach M, Wagner GP. Repressor domain and nuclear localization signal of the murine Hoxa-11 protein are located in the homeodomain: no evidence for role of poly alanine stretches in transcriptional repression. J Exp Zoolog B Mol Dev Evol 2005; 304:468-75.
46. Chariot A, Gielen J, Merville MP et al. The homeodomain-containing proteins: an update on their interacting partners. Biochem Pharmacol 1999; 58:1851-7.
47. Chariot A, van Lint C, Chapelier M et al. CBP and histone deacetylase inhibition enhance the transactivation potential of the HOXB7 homeodomain-containing protein. Oncogene 1999; 18:4007-14.
48. Saleh M, Rambaldi I, Yang XJ et al. Cell signaling switches HOX-PBX complexes from repressors to activators of transcription mediated by histone deacetylases and histone acetyltransferases. Mol Cell Biol 2000; 20:8623-33.
49. Shen W, Chrobak D, Krishnan K et al. HOXB6 protein is bound to CREB-binding protein and represses globin expression in a DNA binding-dependent, PBX interaction-independent process. J Biol Chem 2004; 279:39895-904.
50. Shen WF, Krishnan K, Lawrence HJ et al. The Hox homeodomain proteins block CBP histone acetyltransferase activity. Mol Cell Biol 2001; 21:7509-22.
51. Zhu AH, Kuziora MA. Functional domains in the Deformed protein. Development 1996; 122:1577-1587.
52. Zhu A, Kuziora MA. Homeodomain interaction with the beta subunit of the general transcription factor TFIIE. J Biol Chem 1996; 271:20993-6.

53. Prevot D et al. The leukemia-associated protein Btg1 and the p53-regulated protein Btg2 interact with the homeoprotein Hoxb9 and enhance its transcriptional activation. J Biol Chem 2000; 275:147-53.
54. Yang X et al decapentaplegic is a direct target of dTcf repression in the Drosophila visceral mesoderm. Development 2000; 127:3695-702.
55. Plaza S et al. Cross-regulatory protein-protein interactions between Hox and Pax transcription factors. Proc Natl Acad Sci USA 2008; 105:13439-44.
56. Luo L, Yang X, Takihara Y et al. The cell-cycle regulator geminin inhibits Hox function through direct and Polycomb-mediated interactions. Nature 2004; 427:749-53.
57. Kataoka K, Yoshitomo-Nakagawa K, Shioda S et al. A set of Hox proteins interact with the Maf oncoprotein to inhibit its DNA binding, transactivation and transforming activities. J Biol Chem 2001; 276:819-26.
58. Kirito K, Fox N, Kaushansky K. Thrombopoietin induces HOXA9 nuclear transport in immature hematopoietic cells: potential mechanism by which the hormone favorably affects hematopoietic stem cells. Mol Cell Biol 2004; 24:6751-62.
59. Joliot A, Pernelle C, Deagostini-Bazin H et al. Antennapedia homeobox peptide regulates neural morphogenesis. Proc Natl Acad Sci USA 1991; 88:1864-8.
60. Perez F et al. Antennapedia homeobox as a signal for the cellular internalization and nuclear addressing of a small exogenous peptide. J Cell Sci 1992; 102(Pt 4):717-22.
61. Chatelin L, Volovitch M, Joliot AH et al. Transcription factor Hoxa-5 is taken up by cells in culture and conveyed to their nuclei. Mech Dev 1996; 55:111-7.
62. Derossi D, Joliot AH, Chassaing G et al. The third helix of the Antennapedia homeodomain translocates through biological membranes. J Biol Chem 1994; 269:10444-50.
63. Sugiyama S, Di Nardo AA, Aizawa S et al. Experience-dependent transfer of Otx2 homeoprotein into the visual cortex activates postnatal plasticity. Cell 2008; 134:508-20.
64. In der Rieden PM, Mainguy G, Woltering JM et al. Homeodomain to hexapeptide or PBC-interaction-domain distance: size apparently matters. Trends Genet 2004; 20:76-9.
65. Piper DE, Batchelor AH, Chang CP et al. Structure of a HoxB1-Pbx1 heterodimer bound to DNA: role of the hexapeptide and a fourth homeodomain helix in complex formation. Cell 1999; 96:587-97.
66. Bürglin TR. Analysis of TALE superclass homeobox genes (MEIS, PBC, KNOX, Iroquois, TGIF) reveals a novel domain conserved between plants and animals. Nucleic Acids Res 1997; 25:4173-4180.
67. Mukherjee K, Burglin TR. Comprehensive analysis of animal TALE homeobox genes: new conserved motifs and cases of accelerated evolution. J Mol Evol 2007; 65:137-53.
68. Peifer M, Wieschaus E. Mutations in the Drosophila gene extradenticle affect the way specific homeo domain proteins regulate segmental identity. Genes Dev 1990; 4:1209-23.
69. Nourse J, Mellentin JD, Galili N et al. Chromosomal translocation t(1; 19) results in the synthesis of a homeobox fusion mRNA that codes for a potential chimeric transcription factor. Cell 1990; 60:535-545.
70. Sanchez M, Jennings PA, Murre C. Conformational changes induced in Hoxb-8/Pbx-1 heterodimers in solution and upon interaction with specific DNA. Mol Cell Biol 1997; 17:5369-76.
71. Phelan ML, Rambaldi I, Featherstone MS. Cooperative interactions between HOX and PBX proteins mediated by a conserved peptide motif. Mol Cell Biol 1995; 15:3989-3997.
72. Knoepfler PS, Kamps MP. The pentapeptide motif of Hox proteins is required for cooperative DNA binding with Pbx1, physically contacts Pbx1 and enhances DNA binding by Pbx1 Mol Cell Biol 1995; 15:5811-5819.
73. Johnson FJ, Parker E, Krasnow MA. Extradenticle protein is a selective cofactor for the Drosophila homeotics: Role of the homeodomain and YPWM amino acid motif in the interaction. Proc Nat Acad Sci USA 1995; 92:739-743.
74. Passner JM, Ryoo HD, Shen L et al. Structure of a DNA-bound Ultrabithorax-Extradenticle homeodomain complex. Nature 1999; 397:714-9.
75. Chan S-K, Mann RS. A structural model for a homeotic protein-Extradenticle-DNA complex accounts for the choice of HOX protein in the heterodimer. Proc Natl Acad Sci USA 1996; 93:5223-5228.
76. Shen WF, Chang CP, Rozenfeld S et al. Hox homeodomain proteins exhibit selective complex stabilities with Pbx and DNA. Nucleic Acids Res 1996; 24:898-906.
77. Wilson DS, Sheng GJ, Jun S et al. Conservation and diversification in homeodomain-DNA interactions: A comparative genetic analysis. Proc Natl Acad Sci USA 1996; 93:6886-6891.
78. Wilson DS, Desplan C. Structural basis of Hox specificity. Nat Struct Biol 1999; 6:297-300.
79. Ryoo HD, Marty T, Casares F et al. Regulation of Hox target genes by a DNA bound Homothorax/Hox/Extradenticle complex. Development 1999; 126:5137-48.
80. Ryoo HD, Mann RS. The control of trunk Hox specificity and activity by Extradenticle. Genes Dev 1999; 13:1704-16.

81. Chan SK, Ryoo HD, Gould A et al. Switching the in vivo specificity of a minimal Hox-responsive element. Development 1997; 124:2007-2014.
82. Morgan R, In der Rieden P, Hooiveld MH et al. Identifying HOX paralog groups by the PBX-binding region. Trends Genet 2000; 16:66-7.
83. Ebner A, Cabernard C, Affolter M et al. Recognition of distinct target sites by a unique Labial/Extradenticle/Homothorax complex. Development 2005; 132:1591-600.
84. Li-Kroeger D, Witt LM, Grimes HL et al. Hox and senseless antagonism functions as a molecular switch to regulate EGF secretion in the Drosophila PNS. Dev Cell 2008; 15:298-308.
85. Chan S-K, Pöpperl H, Krumlauf R et al. An extradenticle-induced conformational change in a HOX protein overcomes an inhibitory function of the conserved hexapeptide motif. EMBO J 1996; 15:2476-2487.
86. Lohr U, Yussa M, Pick L. Drosophila fushi tarazu. a gene on the border of homeotic function. Curr Biol 2001; 11:1403-12.
87. Lohr U, Pick L. Cofactor-interaction motifs and the cooption of a homeotic Hox protein into the segmentation pathway of Drosophila melanogaster. Curr Biol 2005; 15:643-9.
88. Galant R, Walsh CM, Carroll SB. Hox repression of a target gene: extradenticle-independent, additive action through multiple monomer binding sites. Development 2002; 129:3115-26.
89. Medina-Martinez O, Ramirez-Solis R. In vivo mutagenesis of the Hoxb8 hexapeptide domain leads to dominant homeotic transformations that mimic the loss-of-function mutations in genes of the Hoxb cluster. Dev Biol 2003; 264:77-90.
90. Merabet S, Kambris Z, Capovilla M et al. The hexapeptide and linker regions of the AbdA Hox protein regulate its activating and repressive functions. Dev Cell 2003; 4:761-8.
91. Merabet S, Saadaoui M, Sambrani N et al. A unique Extradenticle recruitment mode in the Drosophila Hox protein Ultrabithorax. Proc Natl Acad Sci USA 2007; 104:16946-51.
92. Prince F, Katsuyama T, Oshima Y et al. The YPWM motif links Antennapedia to the basal transcriptional machinery. Development 2008; 135:1669-79.
93. Gebelein B, Culi J, Ryoo HD et al. Specificity of Distalless repression and limb primordia development by abdominal Hox proteins. Dev Cell 2002; 3:487-98.
94. Chan SK, Jaffe L, Capovilla M et al. The DNA binding specificity of Ultrabithorax is modulated by cooperative interactions with extradenticle, another homeoprotein. Cell 1994; 78:603-15.
95. Chan SK, Popperl H, Krumlauf R et al. An extradenticle-induced conformational change in a HOX protein overcomes an inhibitory function of the conserved hexapeptide motif. EMBO J 1996; 15:2476-87.
96. Qian YQ, Otting G, Furukubo-Tokunaga K et al. NMR structure determination reveals that the homeodomain is connected through a flexible linker to the main body in the Drosophila Antennapedia protein. Proc Natl Acad Sci USA 1992; 89:10738-10742.
97. Johnson FB, Parker E, Krasnow MA. Extradenticle protein is a selective cofactor for the Drosophila homeotics: Role of the homeodomain and YPWM amino acid motif in the interaction. Proc Natl Acad Sci USA 1995; 92:739-743.
98. Bomze HM, Lopez AJ. Evolutionary conservation of the structure and expression of alternatively spliced Ultrabithorax isoforms from Drosophila. Genetics 1994; 136:965-77.
99. Lopez AJ, Hogness DS. Immunochemical dissection of the Ultrabithorax homeoprotein family in Drosophila melanogaster. Proc Natl Acad Sci Usa 1991; 88:9924-9928.
100. Subramaniam V, Bomze HM, Lopez AJ. Functional differences between Ultrabithorax protein isoforms in Drosophila melanogaster: evidence from elimination, substitution and ectopic expression of specific isoforms. Genetics 1994; 136:979-91.
101. Yaron Y, McAdara JK, Lynch M et al. Identification of novel functional regions important for the activity of HOXB7 in mammalian cells. J Immunol 2001; 166:5058-67.
102. Zhao JJG, Lazzarini RA, Pick L. Functional dissection of the mouse Hox-a5 gene. EMBO J 1996; 15:1313-1322.
103. Vigano MA, Di Rocco G, Zappavigna V et al. Definition of the transcriptional activation domains of three human HOX proteins depends on the DNA-binding context. Mol Cell Biol 1998; 18:6201-12.
104. Tour E, Hittinger CT, McGinnis W. Evolutionarily conserved domains required for activation and repression functions of the Drosophila Hox protein Ultrabithorax. Development 2005; 132:5271-81.
105. Rambaldi I, Kovacs EN, Featherstone MS. A proline-rich transcriptional activation domain in murine HOXD-4 (HOX- 4.2). Nucleic Acids Res 1994; 22:376-82.
106. Ronshaugen M, McGinnis N, McGinnis W. Hox protein mutation and macroevolution of the insect body plan. Nature 2002; 415:914-7.
107. Galant R, Carroll SB. Evolution of a transcriptional repression domain in an insect Hox protein. Nature 2002; 415:910-3.

108. Jaffe L, Ryoo HD, Mann RS. A role for phosphorylation by casein kinase II in modulating Antennapedia activity in Drosophila. Genes Dev 1997; 11:1327-40.
109. Taghli-Lamallem O, Hsia C, Ronshaugen M et al. Context-dependent regulation of Hox protein functions by CK2 phosphorylation sites. Dev Genes Evol 2008; 218:321-32.
110. Zhao X, Sun M, Zhao J et al. Mutations in HOXD13 underlie syndactyly type V and a novel brachydactyly-syndactyly syndrome. Am J Hum Genet 2007; 80:361-71.
111. Utsch B, Becker K, Brock D et al. A novel stable polyalanine [poly(A)] expansion in the HOXA13 gene associated with hand-foot-genital syndrome: proper function of poly(A)-harbouring transcription factors depends on a critical repeat length? Hum Genet 2002; 110:488-94.
112. Greer JM, Puetz J, Thomas KR et al. Maintenance of functional equivalence during paralogous Hox gene evolution. Nature 2000; 403:661-5.

Chapter 2

Cis-Regulation in the *Drosophila* Bithorax Complex

Robert K. Maeda and François Karch*

Abstract

The discovery of the first homeotic mutation by Calvin Bridges in 1915[1] profoundly influenced the way we think about developmental processes. Although many mutations modify or deform morphological structures, homeotic mutations cause a spectacular phenotype in which a morphological structure develops like a copy of a structure that is normally found elsewhere on an organism's body plan. This is best illustrated in *Drosophila* where homeotic mutations were first discovered. For example, *Antennapedia* mutants have legs developing on their head instead of antennae. Because a mutation in a single gene creates such complete structures, homeotic genes were proposed to be key "selector genes" regulating the initiation of a developmental program.[2] According to this model, once a specific developmental program is initiated (i.e., antenna or leg), it can be executed by downstream "realizator genes" independent of its location along the body axis. Consistent with this idea, homeotic genes have been shown to encode transcription factor proteins that control the activity of the many downstream targets to "realize" a developmental program. Here, we will review the first and perhaps, best characterized homeotic complex, the Bithorax Complex (BX-C).

Genetics of the Bithorax Complex: The Model of Ed Lewis

The tale of the Bithorax Complex begins in 1915, when Calvin Bridges[1] discovered the first homeotic mutation, which he named *bithorax* (*bx*). Like all insects, *Drosophilae* have three thoracic segments (T1, T2 and T3). The landmarks of these thoracic segments are pairs of legs emanating from each thoracic segment, a pair of wings that develop from the dorsal part of T2 and a pair of flight organs, called halteres, that develop from T3. In homozygous *bx* mutants, the anterior part of T3 develops like a copy of the anterior part of T2. This phenotype is visible on the body of the adult fly as a transformation of the anterior haltere into anterior wing (Fig. 1).[1] In 1919, Bridges found a second homeotic mutation resembling *bx*. Because of its similarity to *bx*, Bridges named this second mutation *bithoraxoid* (*bxd*). Surprisingly, although *bx* and *bxd* displayed similar phenotypes and mapped to the same chromosomal location, Bridges and Morgan (1923) found that the two mutations complemented and therefore represented different genetic loci.[1] Later, a third mutation with a similar phenotype was found by W.F. Hollander in 1934.[3] In this case, the mutation was dominant with the heterozygous flies harboring swollen halteres (indicating a transformation of haltere (T3) towards wing (T2)). Although through the years, this mutation had been given several names, it acquired its definitive name, *Ultrabithorax* (*Ubx*), in 1950. Unlike the viable *bx* or *bxd* mutations, *Ubx* homozygotes die as first instar larvae.

*Corresponding Author: François Karch—NCCR Frontiers in Genetics and Department of zoology and animal biology. University of Geneva, 30 quai E. Ansermet, 1211 Geneva-4, Switzerland. Email: francois.karch@unige.ch

Hox Genes: Studies from the 20th to the 21st Century, edited by Jean S. Deutsch.

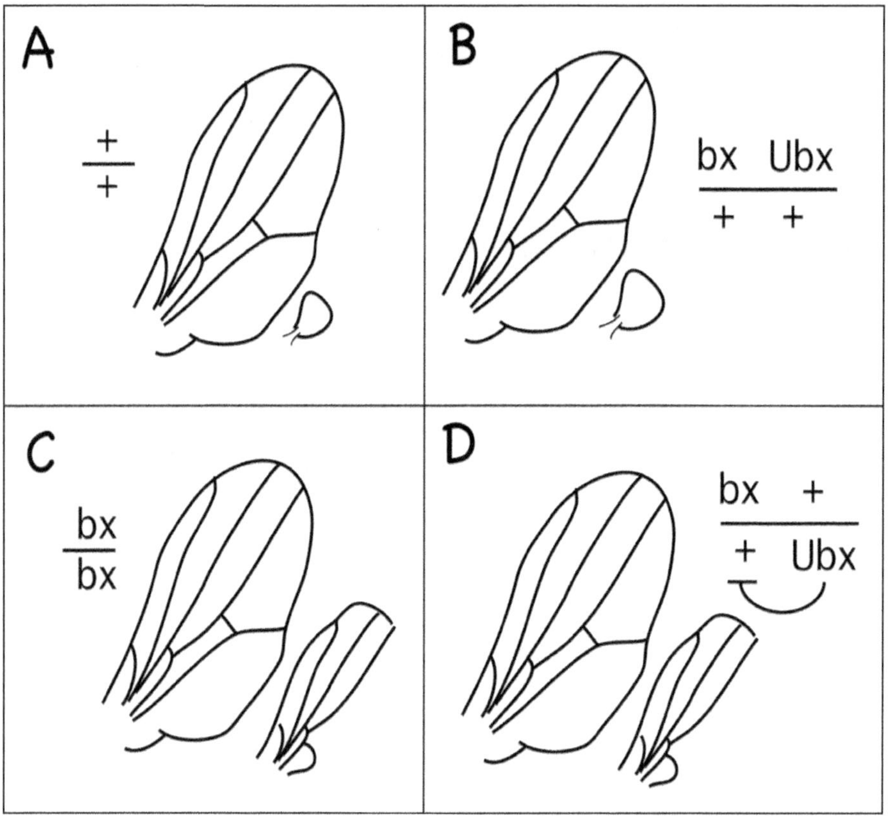

Figure 1. *bx, Ubx* phenotypes and the *cis trans* test. Drawings of wings and halters in wild type flies (A) and in various mutant backgrounds (B,C and D). In *bx* homozygotes, the anterior part of T3 is transformed into the anterior part of T2 as drawn in panel C, where the anterior part of the halter looks like anterior wing. Panel B shows the phenotype of a heterozygote fly carrying a chromosome with a *bx* and *Ubx* mutations. The halter is enlarged to the extend as in a *Ubx* heterozygous fly. Panel D shows a fly *trans*-heterozygous for the same *bx* and *Ubx* mutations (as in B). These flies harbor the same homeotic transformations as observed in *bx* homozygotes (panel C). In other words the *Ubx* mutations appear to inactivate in *cis* the adjacent bx^{+} function. Note that the genotypes in B and D are identical with the differences that the mutations are in *cis* or *trans*. This was the first *cis-trans* test reported in 1951 [6 years before the famous fine structure analysis of the *rII* region of phage T4 by Benzer (1957)[122]].

Interestingly, although the *bx* and *bxd* mutations complement each other, the *Ubx* mutation fails to complement both *bx* and *bxd* mutations. In other words, *Ubx/bx* animals look similar to *bx* homozygous flies and *Ubx/bxd* animals are similar to *bxd* homozygous flies (see Fig. 1, the *cis/trans* figure). This kind of complex genetic interaction where two or more genes appear to occupy the same locus under certain conditions and different loci under other conditions is called pseudoallelism. It was the concept of pseudoallelism that led geneticist Ed Lewis to study these three alleles and what would later be called the Bithorax Complex (BX-C). For this work, Ed Lewis would be awarded the Nobel prize in medicine in 1995 (for details, see two excellent perspectives written in 2002 by Duncan and Montgomery in Genetics[4,5]).

One of the key findings that helped Lewis unlock the mysteries of the Bithorax Complex happened in 1973, when Lewis recovered a complete deletion of the bithorax cluster. Larvae homozygous for this deficiency died at the first instar larvae stage. But what was truly informative about this deficiency was that these larvae died with T3 and all eight abdominal segments (A1 to A8) developing like copies of T2. This phenotype indicated that the bithorax locus, not only contained "genes" specifying T3, but also contained other genes responsible for the identity of each of the abdominal segments. By using this deficiency and a series of smaller deficiencies, Lewis was able to identify the "genes" specifying each of the segments from T3 to A8. Although here, we have used the term "genes", we should note that it was not entirely clear to Ed Lewis if this collection of pseudoalleles really defined genes. Thus, he coined the term "segment-specific-function" to refer to the elements of this allelic series. In his seminal 1978 paper,[6] Ed Lewis described these segment-specific functions for the first time and gave the genetic complex its definitive name, the Bithorax Complex (BX-C).

The actual names of the segment-specific functions described by Lewis are: *abx/bx*, *bxd/pbx* and *iab-2* through *iab-8*.[6] Phenotypic analysis indicated that each class of mutation defined an element that was required for the identity of a single segment. Remarkably enough, these elements mapped to the chromosomes in an order that corresponded to the body segment in which they acted. This correspondence between body axis and genomic organization is referred to as "colinearity" (see Figs. 2 and 3).

Because embryos homozygous for a deficiency of the whole BX-C have all their segments posterior to T2 developing like a copy of T2, Lewis proposed that the second thoracic segment represented the ground state of development (i.e., the default state) and that segment-specific functions allow more-posterior segments to differentiate away from this ground state. Furthermore, the fact that mutations in individual segment-specific functions always caused homeotic transformations towards the last unaffected more-anterior segment (and not always to T2), meant that everything required for more-anterior segment development had to be present in more-posterior segments. For example, in *iab-2* mutant flies, the second abdominal segment (A2) is transformed into a copy of the first abdominal segment (A1) and not into T2. The fact that A2 is not transformed towards the ground state, but into A1 indicates that all the segment-specific functions responsible for differentiating A1 (*abx/bx* for T2->T3) and *bxd/pbx* for T3->A1) respectively) must be functioning in the presumptive A2. Based on these findings, Ed Lewis created a model in which segment-specific functions work in a cumulative fashion to progressively differentiate segments away from the ground state (Fig. 2). Therefore, as one moves to more and more posterior segments, one finds more segment-specific functions in an active state. This additive model is supported by the fact that some mutation that affected anterior segments also caused slight changes in more posterior segments. For example, although the major effect of mutations in the *bxd/pbx* region is to transform A1 into T3, the identities of the more posterior segments are also affected. This is visible by the presence of ventral pits, a feature normally found only on the thoracic segments, on all the abdominal segments. Because of the peculiar colinearity phenomenon described above, Ed Lewis envisioned the BX-C working as genes opening along the chromosome in a segmentally regulated fashion from anterior to posterior (Fig. 2).

The BX-C Encodes Only Three Genes, *Ubx*, *abd-A* and *Abd-B*

Molecular studies of the BX-C began with a textbook example of positional cloning by chromosomal walking.[7-9] As nearly all the mutations affecting the segment-specific functions were associated with chromosomal rearrangement breaks such as inversions, translocations, deficiencies or insertions of transposons, it was possible to quickly localize them on a DNA map by simple, whole-genome Southern. The lesions associated with each class of segment-specific-function always clustered in a discrete part of the BX-C map and the localizations of the numerous mutations (more than 100 have been mapped) confirmed the remarkable colinearity between body axis and order of the segment-specific functions along the BX-C map. The fact that all the mutations in segment-specific functions were due to rearrangement breakpoints also indicated that these lesions did not simply inactivate protein-coding genes (otherwise

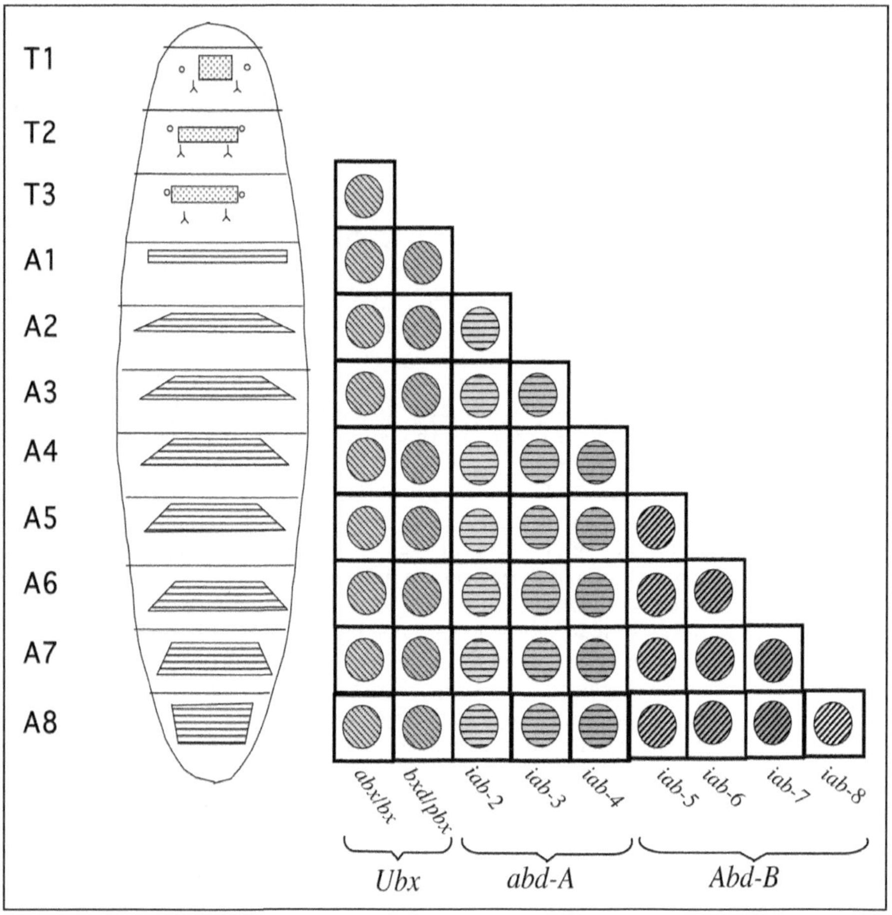

Figure 2. The model of Ed Lewis. A 1st instar larva is drawn along the y axis with symbolized features specific from the 3 thoracic segments (T1-T3) and 8 abdominal segments A1-A8. The orderly segments-specific functions of the BX-C (*abx/bx, bxd/pbx, iab-2* through *iab-8*) are indicated along the x axis. The horizontal brackets indicate which of the segment-specific mutations fail to be complemented by the respective *Ubx*, *abd-A* and *Abd-B* lethal mutations. The dashed circles dots indicate in which segments each of the segment-specific functions are active. Orientations of the dashed patterns (descending, horizontals and ascending) follow the same logic as in Figures 3 and 4. A color version of this figure is available at www.landesbioscience.com/curie.

point mutations would have been also recovered during the numerous screens that had been performed by Ed. Lewis). But if these mutations did not affect regular protein-coding genes (as already suggested by their pseudo-allelism nature), where were the true genes and how many of them existed?

It is worthwhile mentioning that all the mutations in the segment-specific functions are viable as homozygotes. However, other mutations were found in the BX-C that affect more than one segment and are lethal. The *Ubx* mutations, mentioned above, represent one such class of lethal mutations. In 1985, two laboratories (Gines Morata in Madrid and of Robert Whittle in Sussex) described the results of large screens aimed at recovering mutations that fail to complement the

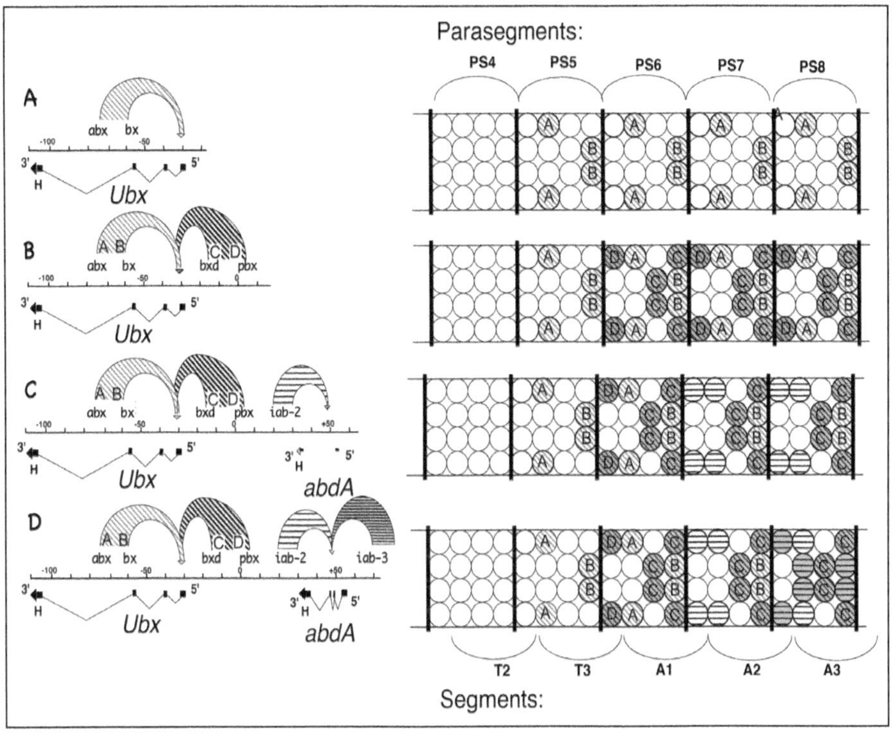

Figure 3. Large orderly parasegment-specific enhancers. The diagrams on the left show successively larger portions of the BX-C with the *Ubx* and *abd-A* transcription units marked, as well of their respective parasegment-specific regulatory units. On the right are cartoons of 5 parasegments (PS4-8) with circles representing individual cells of the ectoderm and nervous system. The corresponding segmental boundaries are indicated at the bottom. Circle filled with descending dashes represent cells expressing *Ubx* and circles with horizontal dashes indicate cells expressing *abd-A*. Row A shows the *Ubx* expression pattern in an embryo that contains only the leftmost 80kb of the BX-C: *Ubx* appears in a few cells of PS5. In the absence of any further BX-C sequences to the right of the *Ubx* promoter, this PS5-specific expression pattern is reiterated in PS6 to PS13. Note that we distinguish the *abx* enhancers that activate *Ubx* in A cells, and the *bx* enhancers activating *Ubx* in B cells. Row B shows the *Ubx* expression pattern in an embryo that contains the leftmost 110 kb of the BX-C, including the *bxd/pbx cis*-regulatory regions. The PS5 expression pattern is unchanged (relative to row A). However more cells express *Ubx* at higher level in PS6 under the *bxd* regulatory elements (C cells) and the *pbx* regulatory element (D cells). Again, without further BX-C sequences on the right of the *bxd/pbx* regulatory region, this PS6-specific expression pattern is reiterated in PS7 to PS13. Row C, same as in row B, but now the embryo contains the left part of the BX-C that include the *abd-A* transcription unit. The Abd-A protein appears in a few cells at low level in PS7 and this pattern is reiterated in the more posterior parasegments (circles filled with horizontal dashes). Note that *abd-A* is turned on in the A and D cells of PS7, shutting off *Ubx* expression (negative *trans*-regulation). Finally, row D shows an embryo containing the left part of the BX-C including *iab-3*. A few more cells express *abd-A* from PS8, at a higher level. Note that *Ubx* is also shut off in the 2 posterior A cells. For simplicity, the *Abd-B* gene is not shown here. *Abd-B* appears in a few cells in PS10 under the control of the *iab-5* regulatory region. At the anterior side, in PS4, the *Antennapedia* gene is expressed. A color version of this figure is available at www.landesbioscience.com/curie.

lethality of the BX-C deficiency.[10,11] These screens identified three complementation groups, each giving rise to homozygous lethality. While one of these corresponded to the already known *Ubx* mutation, the two others were new. The first, *abdominal-A* (*abd-A*), affected the identities of segments A2 through A4 while the second, *Abdominal-B* (*Abd-B*) affected the identities of segments A5 to A8. It is interesting to note that the segmental transformations observed in each these three lethal complementation groups appear as a cumulative effect of the segment-specific functions that they fail to complement. For instance, in *Ubx* mutant embryos, T3 and A1 are transformed into a copy of T2. If such embryos were able to develop, they would give rise to a fly with 3 pairs of wings, the expected phenotype for a triple *bx, pbx* and *bxd* mutant fly. As seen above, *Ubx* mutations fail to complement the *bx, pbx* and *bxd* mutations. Similarly, the segmental transformations observed in embryos homozygous for *abd-A* alleles appear as if the *iab-2, iab-3* and *iab-4* segment-specific functions were affected. This is corroborated by the failure of *abd-A* alleles to complement *iab-2, iab-3* and *iab-4* mutations. And finally, in a similar fashion, *Abd-B* alleles appear to inactivate the *iab-5, iab-6, iab-7* and *iab-8* segment-specific functions. Due to this cumulative effect, it was perhaps not too surprising to find that a *Ubx, abd-A, Abd-B* triple mutant embryo displayed the same segmental transformations as an embryo carrying a complete deletion of the BX-C.[12]

The finding of only three lethal complementation groups suggested that the whole BX-C contained only three genes: *Ubx, abd-A* and *Abd-B*. Molecular studies performed in parallel to the genetic screens confirmed these results. In the laboratory of David Hogness, developmental Northern blots were scanned with overlapping probes from the BX-C to identify transcripts. By mid 1983, a cDNA that spans a 70kb region of the DNA was isolated. This 70kb span of DNA corresponded to the genomic region associated with *Ubx* mutations. Soon after, this cDNA was definitively identified as the *Ubx* gene product.[13] At the same time, cloning of the Antennapedia Complex (ANT-C) by the Gehring and Kaufman laboratories[14,15] led to the identification of the *Antp* transcription unit (covering 100kb of DNA). It was not long before sequence comparisons between the two genes revealed a region of DNA similar in both molecules. This sequence became known as the homeobox.[16,17] The discovery of the homeobox accelerated the identification of the remaining *Drosophila* Hox genes. Very quickly, two other homeobox genes were identified within the BX-C in the regions where the *abd-A* and *Abd-B* mutations had been mapped.[18]

The Segment-Specific Functions Act as Segment/ Parasegment-Specific Enhancers

Genetic and molecular analysis now pointed to the idea that the whole BX-C encoded only three genes. But then, what are the segment specific functions? The description of the expression patterns of *Ubx, abd-A* and *Abd-B* in wild-type embryos and mutant embryos finally led to an answer.[19-23]

In wild-type embryos, *Ubx, abd-A* and *Abd-B* are expressed in broad and overlapping domains with different anterior boundaries of expression. The anterior border of each domain of expression reflects the order of the *Ubx, abd-A* and *Abd-B* genes on the chromosome:[24] *Ubx* is expressed from parasegment 5 to 12 (PS5-12), *abd-A* is expressed from PS7 to PS12 and *Abd-B* from PS10 to 14. Notice that the description of the expression pattern is given in parasegmental units instead of segmental units. This is because within each domain, the pattern of expression of each of the BX-C homeotic genes seems to be made up of reiterated units of expression. These units are approximately the length of a segment, but shifted relative to the normal segmental boundaries and are called parasegments[25] (PS). In the thorax and the abdomen, this shift is approximately half a segment, meaning that a parasegment comprises the posterior half of one segment and the anterior half of the next. PS6, for example, comprises the posterior of segment T3 and the anterior of segment A1.[a]

[a]By chance, this shift is less visible in the adult animal because the visible portion of the adult abdominal segments corresponds primarily to the anterior portion of the parasegment.

By staining various mutant embryos, it was finally understood that the segment-specific functions corresponded to *cis*-regulatory regions that regulate the expression of *Ubx, abd-A* or *Abd-B* in a parasegment-specific fashion. Mutations in any of the segment-specific regulatory regions altered the expression of its relevant target in a specific parasegment. Figure 3 shows a cartoon of *Ubx* expression in the central nervous system from wild-type and different mutant embryos.[26] In PS5, *Ubx* appears at a relatively low level in a few nuclei. In *abx/bx* mutants, *Ubx* expression is lost in those nuclei of PS5 indicating that *Ubx* expression in these cells is under the control of the *abx/bx cis*-regulatory elements.[23,27-29] Similarly, in PS6, *Ubx* appears at a higher level in nearly all nuclei in the PS. In *bxd/pbx* mutants, this PS6-specific expression pattern is replaced by the pattern normally found in PS5, indicating that the PS6-specific expression pattern is controlled by the *bxd/pbx cis*-regulatory region.[27,30,31] Once again, the fact that the PS6-specific expression pattern is replaced by the PS5-specific expression pattern and not by the ground state PS4 pattern, indicates that the *abx/bx cis*-regulatory elements are functional in PS6. Similar findings can be found for each of the mutations in segment specific function.

The finding that the segment-specific functions correspond to *cis*-regulatory domains helped to explain the phenomenon of pseudoallelism in the BX-C. In Figure 4, the *cis*-regulatory regions of the *BX-C* are schematically detailed. The regulatory regions interacting with the *Ubx* gene are shown in light and dark grays. They include the *abx/bx* and *bxd/pbx* regions that regulate *Ubx* expression in PS5 and PS6 respectively.[19,23,27,b] As mentioned above, *bx* and *bxd* mutations fully complement, but mutations in *Ubx* fail to complement both the *bx* and *bxd* mutations. This can now be explained by the fact that these segment-specific functions are *cis*-regulatory elements that modulate *Ubx* expression. Without *Ubx*, these elements have no function. For example, if we look at the contribution of each chromosome to *Ubx* expression independently, a chromosome carrying a *bx* mutation fails to produce *Ubx* protein in PS5 (where the *bx cis*-regulatory element is normally active), but produces the normal amount of *Ubx* protein in PS6 (where the *bxd/pbx cis*-regulatory element is active). The *Ubx* mutant chromosome in *trans*, however, does not produce a functional Ubx product in PS5 or PS6. The resulting *trans*-heterozygote is therefore $Ubx^{-/-}$ in PS5 but $Ubx^{+/-}$ in PS6. Because segment-specific functions behave as recessive mutations, *bx/Ubx* mutants resemble *bx* mutant flies (Fig. 1).

Initiation and Maintenance Phase in BX-C Regulation

In order for the *cis*-regulatory domains to properly control BX-C homeotic gene expression, they must respond to the spatial information provided to the cell in which they reside, to position themselves along the A-P axis of the embryo. In *Drosophila*, the initial determination of A-P position in the embryo is based on the presence or absence of specific transcription factors (for reviews see, refs. 32-35). These transcription factors are encoded by the maternal, gap and pair-rule genes. Together, these transcription factors subdivide the embryo into 14 parasegments. Each of the 14 parasegments then acquires a specific identity (head, thoracic, abdominal segments, genitalia and analia) based on the expression of the homeotic genes. It is now known that the products of the gap and pair-rule genes interact with elements within the *cis*-regulatory regions to determine the initial expression patterns of *Ubx, abd-A* and *Abd-B*.[36-39] For example, the combination of gap and pair-rule gene products expressed in PS10 allows the *iab-5* (but not

[b]*abx* allele stands for *anterobithorax*. Ed Lewis distinguished these alleles from *bx* because they primarily affect the dorsal part of anterior T3. *bx* mutations, on the other hand, affect anterior part of T3 without affecting the dorsal region. However, both type of enhancers are part of the same regulatory region that is active in PS5 (mostly, anterior T3). A similar distinction can be made for the *bxd* and *pbx* elements that are both active in PS6 but in different regions. *pbx* is mostly active in the anterior part of PS6 (mostly corresponding to posterior T3) while *bxd* is mostly active in the posterior part of PS6 (corresponding to anterior A1 in the adult (see Fig. 3).

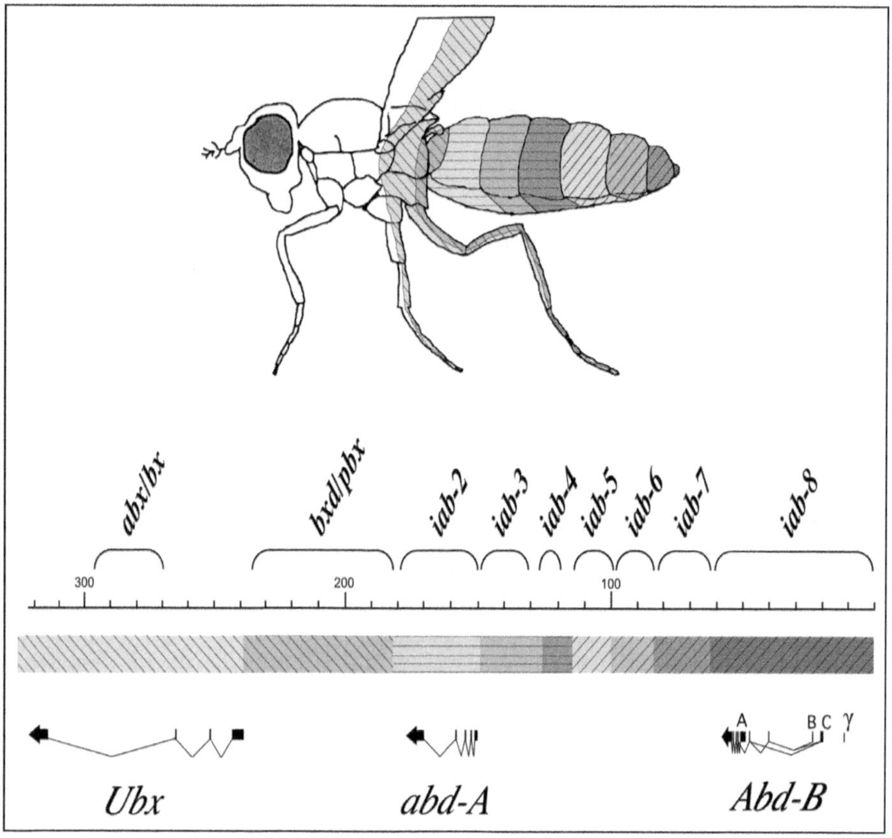

Figure 4. Synopsis of the BX-C. The DNA of the BX-C is depicted as a horizontal bar with different shades of patterns corresponding to the *cis*-regulatory domains. Map coordinate numbering follows the numbering established by the original Drosophila Genome project sequencing of the BX-C (Martin et al, 1995). The three BX-C homeotic genes are indicated below this bar (with exons indicated by the black horizontal bars and the introns indicated by the diagonal lines connecting the bars). The individual *cis*-regulatory domains are indicated by the different shaded patterns. The *cis*-regulatory domains controlling *Ubx* expression (*abx*/bx and *bxd/pbx*) are represented by descending hatched bars. The domains depicted with horizontal lines (*iab-2, 3* and *4*) control *abd-A* expression. Finally, the region represented with ascending hatched bars (*iab-5* through *iab-8*) control *Abd-B* expression. The corresponding adult segments affected by mutations in each *cis*-regulatory region are indicated on the diagram of the fly above the BX-C DNA using the same hatched bars codes. Reproduced from Maeda and Karch, Development 2006; 138:1413-1422, with permission of the Company of Biologists. A color version of this figure is available at www.landesbioscience.com/curie.

the *iab-6*) *cis*-regulatory region to control *Abd-B* expression in PS10/A5. However, this cannot be the sole mechanism to turn on BX-C gene expression because, while the expression of the homeotic genes is set for the whole life cycle of the fly, the gap and pair-rule genes are only transiently expressed during early embryogenesis. Therefore, a memory mechanism is required to maintain the activity state of each of the segment-specific *cis*-regulatory regions after the gap and pair-rule gene products have decayed. The Polycomb-Group (Pc-G) and trithorax-Group (trx-G) genes form the core of this maintenance system.[6,40-43]

While the products of the Pc-G genes function as negative regulators, maintaining the inactive state of the *cis*-regulatory regions not in use, the products of the trx-G genes function as positive regulators, maintaining the active state of the *cis*-regulatory regions activated by early expression of the gap and pair-rule gene products. As many of the products of the Pc-G and trx-G genes have the ability to bind or modify histones, Pc-G and trx-G proteins are thought to maintain homeotic gene expression by modifying the chromatin structure of the *cis*-regulatory domains into active or inactive conformations (for reviews see, refs. 44, 45).

Initiation, Maintenance and Cell Type-Specific Elements within the *Cis*-Regulatory Domain

In order to identify and characterize the various regulatory elements that constitute a *cis*-regulatory region, several laboratories turned to transgenic assays. For these experiments, fragments from the *cis*-regulatory regions were sub-cloned in front of a *lacZ* reporter gene (driven by the *Ubx* promoter) and reintroduced into flies by P-element mediated transformation. From these studies, it was discovered that the *cis*-regulatory regions were modular in nature and composed of three main types of regulatory elements: "initiator" elements, "maintenance" elements (ME) and "tissue-specific/cell type-specific" enhancers.

The existence of the initiators and MEs reflect the distinction between the initiation and maintenance phases of BX-C regulation. By definition, initiator elements confer a parasegmentally restricted expression pattern to the reporter gene during early embryogenesis.[38,46-50] Figure 5 shows the *lacZ* expression pattern driven by one such initiator elements derived from the *bxd/pbx cis*-regulatory regions: the anterior parasegmental boundary of the *lacZ* expression pattern in PS6, corresponds to the parasegment that is specified by the *bxd/pbx cis*-regulatory regions. Dependence on the expression of the gap and pair-rule genes has been shown for some of these initiators using the same *lacZ* reporter gene assays.[47-49,51] We now know that each of the segment-specific *cis*-regulatory regions contain at least one such initiator element.

In most cases, the anterior border of expression of a reporter gene controlled by an initiator element is lost when the products of the gap and pair-rule genes decay (at the end of the initiation phase; see Figure 5 panel D). However, a few larger fragments are able to maintain the initial anterior border of expression of a *lacZ* reporter. The ability to maintain the initial expression pattern has been mapped to fragments of DNA that are distinct from the initiator and are called Maintenance Elements[52] (see panel B of Fig. 5). Because the maintenance of the initial expression pattern of the reporter gene is lost in embryos that are mutant in Pc-G genes, maintenance elements are also referred as to Polycomb-Response-Elements.[53-57] These maintenance elements do not have an intrinsic parasegmental address and can maintain different parasegmental patterns when combined with different initiators Simon et al 1993.[58] Polycomb-Response Elements (PREs) can also impose silencing to other linked reporter genes including the *white*$^+$ gene that is used to score transformant lines.[55,59-61]

Another feature of initiators that should be emphasized is that although these constructs accurately report the PS position from which a *cis*-regulatory region is functional, they do not accurately reflect the cell-specific expression of the *cis*-regulatory domain. In other words, the homeotic genes are not necessarily expressed in the same cells within a PS as the reporter. The cell-type-specific expression pattern is given by a third type of regulatory element found within the BX-C *cis*-regulatory regions.[46,50,62-64] In most cases, these elements confer a cell/tissue-specific expression pattern to a reporter gene that is reiterated in all the parasegments along the A-P axis of the embryo. It must be noted, however, that within the BX-C, tissue specific expression of the homeotic genes is restricted parasegmentally along the A-P axis. This apparent discrepancy indicates that within the context of the *cis*-regulatory regions, the activities of the cell type-specific enhancers are regulated by the initiators and maintenance elements residing within the regulatory region.

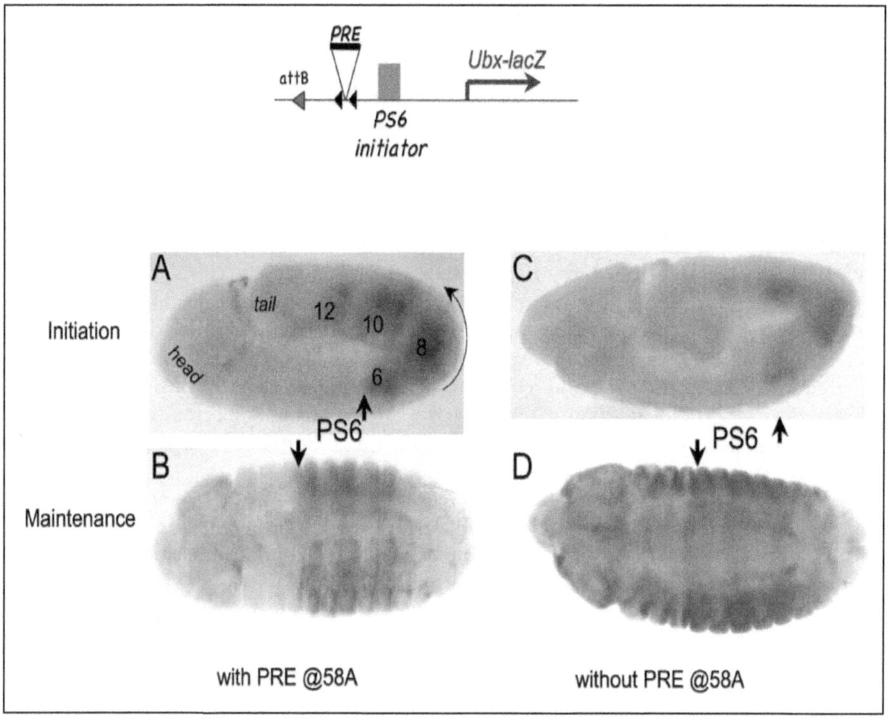

Figure 5. Initiation and maintenance elements. The figure illustrates the analysis of initiation (PS6 initiator) and maintenance (PRE) elements with the reporter gene construct depicted above, and that includes a *lacZ* gene driven by the *Ubx* promoter. Note that the constructs are introduced into flies with the site-specific integrations system mediated by the φC31 integrase, enabling integration of the transgenes in a defined chromosomal context (58A in this example). In addition the PRE is flanked by *lox P* sites, allowing to compare the activity of the initiator in the presence or absence of the maintenance element. Early embryos are shown on top (A and C) during the initiation phase. At this particular stage called extended germband, PS8 to 14 curve around towards the dorsal side (as indicated by the curved arrow). *LacZ* expression (detected with antibodies) appears from PS6 in a pair-rule fashion (PS8, PS10 and PS12). In the presence of the PRE sequence, the anterior border of expression is maintained from PS6 at later stages (panel B). When the PRE is deleted, the initial PS6 anterior border of expression is lost at later embryonic stages where *lacZ* expression invade the anterior parasegments (panel D). A color version of this figure is available at www.landesbioscience.com/curie.

The *Cis*-Regulatory Regions Are Organized in Segment-Specific Chromosomal Domains

How can the various enhancers in a *cis*-regulatory region be coordinately regulated? Three types of observations indicate that the *cis*-regulatory regions are in fact organized into parasegmenally regulated chromosomal domains. All three arguments are genetic in nature.

Thus far, we have restricted our description of mutations to loss-of-function mutations (LOF) that transform a given parasegment into a copy of the parasegment immediately anterior to it. However, there are also dominant gain-of-function (GOF) mutations that cause the opposite transformations, where a given parasegment is transformed into a copy of the parasegment immediately posterior to it. In other words, the GOF alleles appear to behave as if a given regulatory region is activated one parasegment anterior to where it should normally be active. These GOF mutations have been very instrumental in shaping the logic of the Lewis model.

The first observation supporting the organization of the regulatory regions into parasegmentally-regulated chromosomal domains comes from one of the most astonishing BX-C alleles, *Contrabithorax*[1] (*Cbx*[1]; see Peifer et al 1987[26]). This dominant mutation, recovered by Ed Lewis in 1949, causes the transformation of the posterior part of T2 into the posterior part of T3. For the sake of clarity, the *Cbx*[1] phenotype is opposite to the transformation observed in *pbx* mutants (*pbx*[1]; see above) where the posterior part of T3 is transformed into the posterior T2. In the logic of the Lewis model, *Cbx*[1] flies look as if the pbx^+ function was activated one segment ahead, in posterior T2.

Fine structure mapping revealed that *Cbx*[1] was actually associated with two complementary DNA lesions within the BX-C. One lesion maps to the right of *bxd* and turns out to be the *pbx*[1] lesion itself (Fig. 3). Meanwhile, the second lesion maps just to the right of *bx* and is associated with the dominant GOF phenotype. In 1983, the cloning of *Cbx*[1] confirmed the genetic mapping data (Bender et al, 1983).[8] In the initial double mutant, a 17kb piece of DNA had been deleted from the *bxd/pbx* regulatory region and reinserted 44kb away within the *abx/bx* region of the BX-C. In 1987, Peifer et al.[26] proposed a model to explain, not only how the *Cbx*[1] mutant works, but also how the whole BX-C might work. According to the model, the 17kb long fragment that is deleted from the *bxd/pbx* regulatory region contains enhancers regulating *Ubx* in specific cells in PS6 to create the posterior part of T3 (the D-cells in Fig. 3). The homozygous deletion of these enhancers would normally lead to the loss of *Ubx* expression in these cells and consequently the transformation of the posterior part of T3 into the posterior part of T2. This is the same model that has been used to explain the LOF mutations in the BX-C (Fig. 3). However, the transposition of these enhancers somehow leads to the more drastic dominant *Cbx* phenotype. This was surprising since enhancers are, in general, able to act on their target promoter independent of their position. Therefore, moving these enhancers from an upstream position to a downstream position relative to the *Ubx* promoter should normally not affect their activity. To explain these findings, Peifer et al[26] proposed that a parasegmental address might be conferred by the DNA domain in which the enhancers reside. In this model, each regulatory region would be embedded in a DNA domain that would be coordinately activated in a parasegment-specific manner. In the *Cbx*[1] mutant, the *pbx* enhancers placed in the *abx/bx* domain are activated in cells that are equivalent to the cells they activate normally in PS6, but in PS5, thereby transforming posterior T2 into posterior T3.

Since this early hypothesis, other evidences have supported the parasegment-specific domain model. One of the strongest evidences comes from enhancer-trap lines within the BX-C. In *Drosophila*, transgenic animals are often made using P-element transposons. These transposons insert throughout the genome in a fairly random fashion. If these P-elements contain a basal promoter and a reporter construct, they can be used to visualize the enhancer elements around an insertion site. This technique is called enhancer trapping.[65] A number of enhancer trap lines have been isolated within the BX-C. For example, the anterior border of *lacZ* expression for three transposons inserted within the *abx/bx cis*-regulatory is PS5. This mimics the pattern of expression given by the *abx/bx* region, which controls *Ubx* expression in PS5. What is amazing about this finding is that the outer two insertions are over 70 kb away from each other. Meanwhile the anterior border of expression of a fourth insertion, located just a few kb further to the right these insertions, is PS6 and mimics the pattern of expression given by the *bxd/pbx cis*-regulatory region. Examining the large number of enhancer trap lines isolated in the BX-C,[66] made two striking observations. First, lines spread out over quite great distances often have the same parasegmental anterior border of expression, while other lines, located just a few kb away display a different anterior border of expression. Second, the anterior border of *lacZ* expression always progresses towards the posterior by increments of one parasegment. Although these lines are trapping different enhancers, lines can be grouped by the parasegment in which they start to express. These observations are in perfect agreement with the initial proposal made by Peifer et al (1987 see above), in which BX-C enhancers reside in chromosomal domains that are coordinately regulated in a parasegment-specific fashion (see ref. 66 for a review).

Chromatin Boundaries Flank the Parasegment-Specific Domains

The third line of evidence supporting the domain hypothesis is the presence of specialized elements called boundary elements. One prediction made by the domain hypothesis is the existence of elements to limit the extent of each domain. Indeed, the sharp parasegmental transition in expression of the enhancer trap lines supports this hypothesis. In Figure 4, the boundaries are symbolized by the sharp color (or gray) transition between the adjacent domains symbolized by the rectangles. The presence of a boundary is postulated between each of the regulatory domains. Thus far, three boundaries, *Mcp*, *Fab-7* and *Fab-8*, have been conclusively identified through molecular and mutational analysis.[50,67-70] All three boundary deletion mutations are associated with a dominant gain-of-function phenotype. The best characterized of them is *Fab-7*, which separates the *iab-6 cis*-regulatory domain from the *iab-7 cis*-regulatory domain. In *Fab-7* mutants, PS11/A6 is transformed towards PS12/A7 identity (posterior oriented transformation). In effect, *iab-7*, which is normally active only in PS12/A7, is activated one parasegment ahead, within PS11/A6. In agreement with this, *Abd-B* expression is regulated in a PS12-like pattern,[c] transforming cell identity from PS11 to PS12.[71,72] Based on these results, it was proposed that in *Fab-7* mutants, the *iab-6* and *iab-7* domains become fused into a single functional unit with mixed characteristics: parasegment <u>specificity</u> being provided by *iab-6* (initiation), while parasegment <u>identity</u> is provided by *iab-7*.[67] In a case very similar to the *Cbx*^{1} example, enhancers from one domain become controlled by the initiator of another domain. Similar findings have been found for the *Mcp* boundary that separates *iab-4* from *iab-5* and the *Fab-8* boundary that separates *iab-7* from *iab-8*.[50,68] Recently, the *Fab-6* boundary was localized via inference from the phenotypes of two relatively large deletions that fuse the *iab-5* and *iab-7* domains.[64]

Elements Mediating Long-Distance *Cis*- and *Trans*- Regulatory Interactions

All of the BX-C parasegment-specific *cis*-regulatory domains have now been mapped based on mutations and enhancer trap lines. Overall, they show that the BX-C *cis*-regulatory domains are quite large (each from 10-30 kb in length) and seem to contain numerous cell-type specific enhancers. What is so surprising about this is that each of these enhancers faithfully regulates their target promoter from distances as far as 60 kb away. How these *cis*-regulatory domains interact with distant promoters is currently a topic of intense investigation.

Transvection Studies

The first evidence of long-distance interactions within the BX-C came with the discovery of transvection. In the early 1950s, when the nature of the gene was still largely unknown, Ed Lewis came up with a model to explain the complex complementation analysis of his bithorax pseudo-allelelic series (see Fig. 6).

This model was based largely on the biochemical ideas of the day, where gene function was thought to be constructed through metabolic enzymatic reactions. The sequential reactions of metabolic pathways seemed to provide a reason for the colinearity of BX-C mutations. Lewis' "sequential reaction model" is depicted, in part, in Figure 6. According to this model, the *bx*$^+$ function is responsible for modifying a substrate S into a substance A. Then, the substance A diffuses in *cis* to become a substrate for the *Bxl*$^+$ enzyme (one of the previous names for *Ubx*$^+$) and is transformed into a substance B. Substance B then diffuses in *cis* to the *bxd* enzyme and is transformed into substance C and so on. In this scheme, substances A or B had the ability to promote a T3 identity from the default T2 fate. Meanwhile, substance C would promote development of A1. A posterior to anterior concentration gradient of substance S was imagined to explain why the BX-C genes are active in T3 and not in T2, where the level of S would be too low to produce substantial amounts of A, B or C. With this sequential reaction model, Lewis was able to explain all the complex complementation data.

[c]*Abd-B* expression is higher and expressed in more cells in PS12 than in PS11.

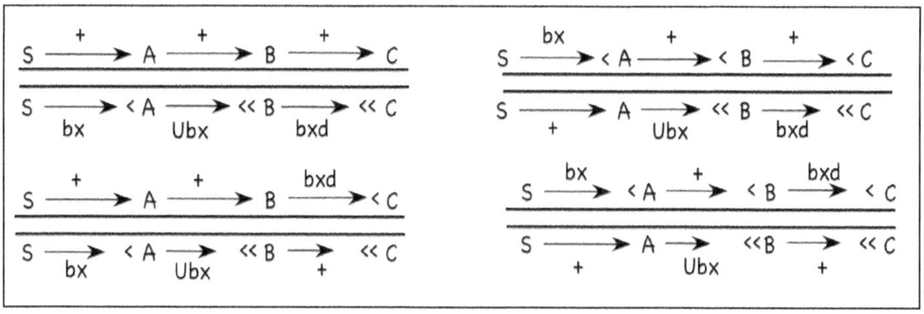

Figure 6. Model of the complex complementation analysis of Lewis's *bithorax* pseudo-allelelic series. Adapted from Lewis (1951).[73]

As science's understanding of the nature of the gene evolved, many scientists questioned Lewis's sequential reaction model. They wanted explanations in terms of missense and nonsense mutations within as single protein-coding unit. As this kind of explanation did not fit Lewis'data, Lewis felt obliged to defend his model and designed a genetic test to support his sequential reaction model. It was through this test that Ed Lewis discovered the phenomenon of transvection.

The idea of local diffusion of enzymatic substrates was a critical point to Lewis' model. Since homologous chromosomes remain intimately paired throughout the cell cycle in *Drosophila* (a feature of all dipterans), Lewis imagined that diffusion might lead to substrates occasionally diffusing to the copy of the gene on the paired homologous chromosome. If so, that meant that mutations on one chromosome might be able to be partially complemented by a functional gene in *trans*. Furthermore, Lewis hypothesized that this complementation would be pairing dependent and could be abolished by chromosomal rearrangements that prevented pairing. Lewis, therefore, began looking for mutant combinations that partly complemented in *trans* in a pairing dependent manner. Remarkably, Lewis identified such complementing alleles and named the phenomenon he discovered, transvection.[74] It is worthwhile emphasizing here that Ed Lewis did not discover transvection by identifying rare events taking place during the numerous and extensive screens he performed. Rather transvection was discovered though hypothesis driven experimentation aimed at proving his sequential reaction model.[d] Although Lewis' mechanistic interpretation of transvection may have been incorrect, his work on transvection led to the understanding that the BX-C *cis*-regulatory elements can work in *trans* and at a distance.

Since Lewis' first experiments, transvection has been studied extensively in the BX-C, as well as at other loci. These loci have in common extensive regulatory regions and it is now believed that transvection reflects the ability of enhancers to activate promoters in *trans*. In very elegant studies at the *yellow* locus, the laboratories of Ting Wu and Pamela Geyer have shown that enhancers are better at activating in *trans*, when their natural *cis* target promoter is physically affected by a mutational lesion.[76,77] Although promoter alteration of one of the alleles may not be a prerequisite for transvection to occur, this observation appears hold true in the BX-C.[78,79]

While transvection at *Ubx* is strictly dependent on homologous chromosomal pairing, three groups discovered that transvection at the *Abd-B* locus follows different rules.[79-81] Transvection of the *iab-5*, *6* and *7* regions on the *Abd-B* gene in *trans* was shown to be quite strong and difficult to disrupt by rearrangement breaks.[78] Hopmann and Duncan argue that the tenacity of transvection at *Abd-B* reflects the fact that the *iab-5*, *6* and *7* regulatory domains must normally act over extremely long distances to interact with the *Abd-B* promoter in *cis* (from 20-60 kb away). Using this transvection assay, they were able to map the first element mediating transvection, called the

[d] It should be noticed that in 1953 Madeleine Gans[75] discovered almost simultaneously a very similar phenomenon in which particular *zeste* alleles caused mis-expression of *white*$^{+}$ in a fashion that depended on homolog pairing of the *white* locus.

"transvection mediating region" (tmr), which is located in a region between the *Fab-8* boundary and the *Abd-B* transcription unit.

A second element mediating transvection in the BX-C was discovered by Sipos et al in 1998.[81] In this work, a different method was used to characterize transvection at *Abd-B*. Instead of following activation by *iab-5, 6* and *7* on an *Abd-B* promoter in *trans*, they observed the weakening of the *Fab-7* GOF phenotype by providing in *trans Abd-B* point mutant alleles. This suppression was shown to be due to a sharing of the *iab-6/iab-7* fused *cis*-regulatory domain between the *cis* and *trans* promoters. Using this transvection model, they discovered a region upstream of the *Abd-B^m* promoter that seems to mediate transvection (called the tethering region). Unlike the other transvection studies in the BX-C, the form of transvection monitored by Sipos et al, was sensitive to pairing.[81]

These transvection studies clearly demonstrated the existence of elements within the BX-C that mediate long distance regulatory interactions. However, these studies were based on alleles found through traditional mutagenesis techniques and therefore could not more precisely define the transvection mediating elements. Because, until recently, targeted mutagenesis was not possible in *Drosophila*, people have used transgenic assays to identify potential long-distance interaction motifs.

Experiments using *lacZ* reporter constructs to identify and test Polycomb-Response Elements (PRE) have shown that PREs may play an important role in mediating some long-distance interactions. PREs have long been known to act as silencers on transgenes. In fact, the marker used to find transgenic insertion events in flies, the *white* gene, is often silenced by PREs present on the transgene. The *white* gene is required to produce color in the eyes of adult flies and therefore *white* mutant flies have white eyes. A transgene carrying a copy of a minimal *white* gene can partially complement a *white* mutation. This is a dose-dependent effect, with flies homozygous for a mini-*white* transgene generally having darker eyes than a heterozygous sibling. However, when a transgene contains a PRE, this relationship is reversed, with homozygotes often displaying a lighter eye color than heterozygous siblings. Like transvections, this synergy for PRE mediated silencing has generally been shown to require the pairing of homologous chromosomes. Because of this pairing requirement, this phenomenon has been called pairing-sensitive silencing (PSS).[55,59-61,82-84] However, like transvection at *Abd-B,* some PREs seem to be able to synergize from distant locations (even on different chromosomes), indicating that they might physically interact.[83,85] Under the microscope, these interactions have been documented, with P-elements containing PREs inserted at different locations localizing in close proximity to one another in interphase nuclei.[86-88] The ability of PRE sequences to find each other within the nucleus is believed to result from the coalescence of individual Polycomb-Repressing complexes and their associated chromatin targets into larger structures referred to as Polycomb bodies.[87,89,90] In agreement with this, P-elements carrying PRE sequences often insert into chromosomal locations that are already regulated by Polycomb-complexes.[91]

While these PRE-harboring P-elements tend to insert into chromosomal locations that are normally associated with Pc-G repressing complexes, they do not specifically reinsert at the sites from which they originate. There are, however, rare cases where a DNA fragment can direct its own insertion into the vicinity of the site from which it originates. This phenomenon is referred as to as "homing" and was first discovered with a fragment overlapping the promoter of the *engrailed* gene and its 5' upstream sequences.[92,93] A second such homing fragment is a 7 kb fragment derived from the region separating *bxd/pbx* from *iab-2* in the BX-C. In this case, 18% of the P-elements containing this homing fragment inserted into the BX-C. This homing fragment was actually used to generate many of the enhancer trap lines recovered in the BX-C (see above).[66] While the mechanisms behind homing remain elusive, it is worthwhile mentioning that the "homing pigeon" fragment spans the boundary region separating the *bxd/pbx* regulatory domain from the *iab-2* domain. The idea of boundaries mediating homing is not without merit. The *bluetail* transposon (*blt*), another of the enhancer trap lines inserted into the BX-C may constitute another case of a boundary-mediated homing event. In the process of studying the

R73 region of the BX-C for enhancer activity, a 1kb-long DNA fragment spanning the *Fab-8* boundary was placed onto a P-element transposon. In the 23 independent insertions of this construct one landed right next to the *Fab-7* boundary.[71] Although this could be coincidental, newer experiments suggest otherwise.

Recently, work from our laboratory has provided perhaps the strongest support for the idea that boundaries mediate long-distance chromatin interactions. Using a modified DamID method, we were able to document an association between the *Fab-7* boundary element and a region near the *Abd-B* promoter.[94] Interestingly, this interaction was only found in anterior tissues (where *Abd-B* is silenced) and not in more posterior tissues (where *Abd-B* is expressed), suggesting that this interaction is regulated along the A-P axis. Furthermore, we showed that this interaction is absolutely dependent on the presence of the *Fab-7* boundary element. Based on these observations and boundary element genetics, a model has been proposed in which BX-C boundaries play an active role in targeting the enhancer regions to their promoters.[94,95]

While PREs and boundaries are able to interact over long distances or even across chromosomes, two additional regulatory sequences of the BX-C have been identified that may help establish enhancer-promoter interactions. These two elements are called the "Promoter-Targeting Sequence" (PTS) and the "tethering element".

Promoter Targeting Sequences

Nearly all chromatin domain boundaries seem to be able to function as chromatin insulators (such as the *scs/scs'*, *gypsy* and *β-globin* 5'HS4 insulators; for reviews see for example.[96] In general, insulator activity is determined by a transgenic assay. DNA fragments suspected of insulator activity are placed between an enhancer and a reporter gene promoter to see if the DNA fragment is able to suppress the expression of the reporter gene. If the fragment suppresses the reporter gene when placed between the enhancer and promoter, but not when placed elsewhere, the fragment is considered to be an insulator.[97,98] Because the activity of the *Mcp*, *Fab-7* and *Fab-8* boundaries are reminiscent of the activity of chromatin insulators, each boundary has been tested for insulator activity. In the transgenic assay described above, each of the BX-C boundary elements (*Mcp*, *Fab-7* and *Fab-8*) has been proven to have insulator activity.[49,50,99-101] However, this finding leads to a paradox. Insulators, by definition, block enhancers from interacting with a target promoter when placed in between these two elements. In the BX-C, boundary elements, like *Fab-7* and *Fab-8*, are located in between many BX-C enhancers and their target promoter. How then can these enhancers ever reach their target promoter over so many intervening insulators? In 1999, Zhou and Levine[102] looked for specific DNA fragments that could aid distal enhancers in bypassing intervening boundaries. The result of these experiments was the identification of an element that they called the promoter targeting sequence (PTS). This fragment was isolated from a region of the BX-C located in the *iab-7* domain, just adjacent to the *Fab-8* boundary. When this fragment is placed next to the *Fab-8* insulator in a common insulator assay, it allows distal enhancers to bypass the *Fab-8* insulator. Moreover, it was shown that this PTS element is also able to aid an enhancer in bypassing other insulators (like the *gypsy* insulator), suggesting that PTS function is independent of the insulator itself.[102] Since these initial discoveries, a second PTS element was found in 2005, in a DNA fragment from the *iab-6* domain.[103] Based on these results, the authors suggest that each BX-C boundary element may be flanked by a PTS element to aid in insulator bypass. Although this is an attractive hypothesis, studies from our lab have complicated this issue. First, using gene conversion, we have replaced the *Fab-7* boundary by the *gypsy* insulator and found that both the *iab-6* and *iab-5* regulatory domains are no longer able to regulate *Abd-B*. Thus, although in transgenic constructs, PTS elements seem to allow enhancers to regulate gene expression across a *gypsy* insulator, in situ, their activity seems to be different. Second, we recently recovered simple deletions of both known PTS elements (in *iab-6* and *iab-7*, respectively). These deletions are not associated with any visible phenotype indicating that if the PTS hypothesis is correct, each domain must contain multiple, redundant PTS elements.[64]

Over the years, the PTS hypothesis has received an inordinate amount of attention. The reason for this is obvious. The BX-C has always been thought of as a complicated network of *cis*-regulatory interactions and the presence of PTS elements simplified the picture substantially. Besides solving the problem of bypassing intervening insulators in the BX-C, it solved the problem of how enhancers find their target promoter over long distances. However, now, with numerous genetic arguments pointing against the simple PTS hypothesis, we must question its validity and the assumptions that led to its discovery. To start with, we must question the idea that insulators in the BX-C create a problem. Insulators are elements defined by transgenic assay. By this assay, insulators seem to act negatively on enhancers or silencers. But mechanistically, what does it mean to have insulator activity? The DamID studies performed in our lab demonstrate that the *Fab-7* boundary is an element required for chromatin-chromatin interactions that bring the *cis*-regulatory domain around *Fab-7* to the *Abd-B* promoter. Assuming all boundaries behave in the same way, we can imagine that these interactions would create loops of chromatin at the *Abd-B* promoter, where each loop would contain a separate *cis*-regulatory domain. But what would be the consequence of putting one of these elements on a transgenic reporter construct? Localization to the BX-C? Abnormal interaction with the promoter? In any case, insulation could be one outcome. Based on the activities that PTS elements have in transgenes, they are probably doing something in the regulation of the BX-C. Their peculiar ability to lock enhancer-promoter interactions hints that PTS elements may play a role in stabilizing enhancer-promoter communication.[104] However, without a solid genetic indication of its role in vivo, the role of PTS elements in homeotic gene regulation is still a mystery.

Promoter Tethering Element

The latest element potentially mediating long-distance interactions is the promoter tethering element, molecularly identified by Akbari et al (2008)[105] Its discovery followed the observation that the IAB5 initiator element (the PS10-specific initiator belonging to *iab-5*[62,64]) is almost equidistant from the two distant promoters, *abd-A* and *Abd-B* but nevertheless regulates only *Abd-B*. Using transgenic reporter gene assays, Akbari et al[105] searched for elements in the vicinity of the *Abd-B* promoter that would be responsible for that specificity. To do this, they created a transgenic reporter construct carrying the IAB5 initiator element and both the *Abd-B* promoter and the *abd-A* promoter controlling the expression of different reporter genes. By systematically deleting parts of the *Abd-B* promoter, they were eventually able to identify a 255 bp element, located 40 bp 5′ from the *Abd-B* transcriptional start site, that is able to mediate IAB5 specificity for the *Abd-B* promoter. This element they named the "tethering element" is located within the tethering region 5′ to the *Abd-B* promoter responsible for transvection.[81] Based on these observations, it seems likely that part of the specific long-range enhancer-promoter interactions in the BX-C are regulated by the interactions between the enhancers and the promoters themselves.

Intergenic Transcription in the BX-C

Transcription of the regulatory domains has been known for almost 25 years. These intergenic transcripts were first detected in examinations of cDNA clones from the *bxd/pbx* regulatory domain.[106] Sequencing of the cDNA clones did not reveal any significant open reading frames. Later, in 1989, using large genomic probes spanning from *iab-2* through *iab-8* on tissue sections from early embryos, Sanchez-Herrero and Akam identified three additional transcribed domains.[107] Interestingly, the anterior limits of expression of these transcripts were colinear with the arrangement of the probes along the chromosome. A fifth transcript was found the following year in the *iab-4 cis* regulatory domain.[108] This transcript is expressed in the early embryo in a broad domain with an anterior limit corresponding to PS8 or PS9. For nearly 10 years, the existence of these transcripts was thought to be a byproduct of domain accessibility, with cryptic promoters in the large regulatory regions firing in the domains free from Polycomb

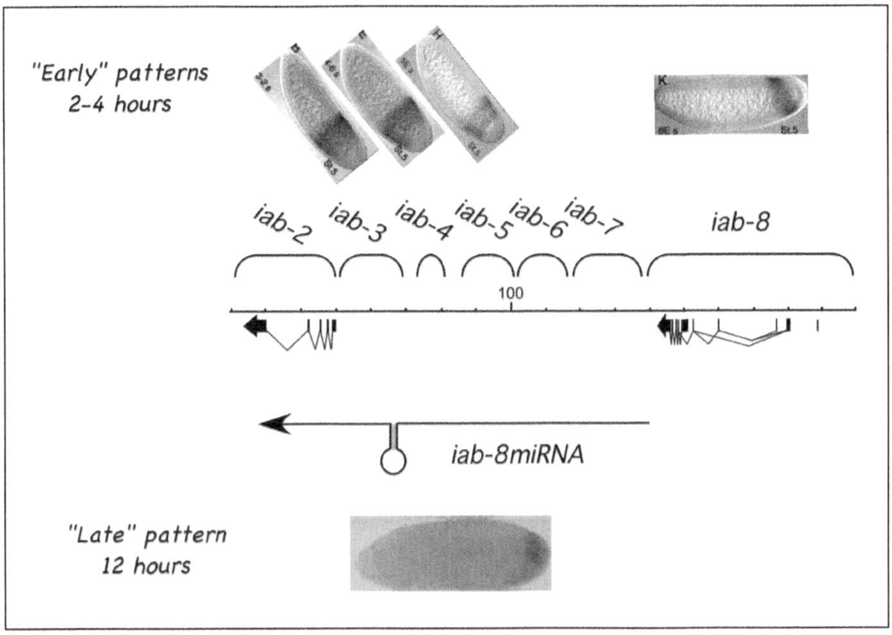

Figure 7. Intergenic transcription in the abdominal part of the BX-C. The horizontal line represents the genomic DNA of the abdominal part of the BX-C with the *abd-A* and *Abd-B* transcription units and their associated *cis*-regulatory regions *iab-2* through *iab-8*. For the sake of simplicity, only sense transcripts are depicted (relative to the coding transcripts of *abd-A* and *Abd-B*). The expression pattern of the early nc RNAs are shown above the genomic DNA with probes derived from the *iab-3, iab-4, iab-5, iab-6* and *iab-8 cis*-regulatory domains [embryos *iab-3, 4, 5* and *8* are derived from Bae and Drewell (2002), while embryo *iab-6* was performed in our laboratory by S. Galetti, (unpublished)]). The expression pattern of the later *iab-8*miRNA is shown below the genomic DNA (Bender; 2008). The structure of the *iab-8*miRNA transcription unit is depicted below the genomic DNA. Note that the 3′end of the transcription is not clearly established (see Bender 2008).[111] A color version of this figure is available at www.landesbioscience.com/curie.

silencing. However, the discovery of RNA interference led to renewed interest in these noncoding RNAs (ncRNAs).

The detailed expression patterns of many of the BX-C intergenic transcripts were determined by Drewell and colleagues.[109,110] There are four major observations worth mentioning. First, with two exceptions, all of the intergenic transcription activity in the BX-C seems to emanate from the sense strand of the BX-C (the same strand that the *Ubx*, *abd-A* and *Abd-B* transcription units emanate from (see Fig. 7). Second, most of these intergenic transcripts appear very early, in stage 5 blastoderm embryos. This period corresponds to the initiation phase of BX-C regulation, when the products of the gap and pair rule genes activate the regulatory domains. Third, the spatial distribution of the intergenic transcripts reflects the activity of the respective *cis*-regulatory domains from which they emanate. For example, probes from the *iab-5* regulatory domain show ncRNA transcripts from PS10 to PS13.[109] And finally, at stage 9, roughly an hour and a half later, the majority of these expression patterns vanish and are replaced by a new pattern that is specific for PS13 and PS14. As probes from *iab-2* through *iab-7* light up a similar expression pattern in PS13 and PS14, it is believed that there is a single, 120kb long transcript, initiating just downstream from the *Abd-B* homeobox and travelling all the way to the *iab-2* regulatory domain.[49,109,111]

Given that the expression of these ncRNAs coincides with the initiation of BX-C gene expression many groups have speculated about possible roles for these transcripts in the process of initiation. One of the most intriguing and controversial of these ideas is that these ncRNA are a method of inactivating PREs.

As described above, the Bithorax Complex is thought to work in two phases, initiation and maintenance. Initiation is controlled by elements called initiators that respond to signals from the gap and pair-rule genes to activate a regulatory domain. Maintenance elements then maintain the active or silenced state through the Polycomb and trithorax groups of proteins. Exactly how initiators communicate with MEs is still unknown. However, three papers published in 2002 showed that the act of transcription over a PRE early in development can inactivate the PRE's silencing capability.[112-114] As the ncRNAs transcribe the *cis*-regulatory domains early, this looked like a possible role for the ncRNAs.

A few years after these initial observations were reported, two conflicting papers were published showing that this issue is far from solved. The first work was performed in the laboratory of Frank Sauer. In this work, Sauer and colleagues identified three ncRNAs from the *bxd* PRE area of the BX-C and presented evidences that these ncRNAs act in an instructive manner to activate *Ubx* expression through directly recruiting the Ash1 (trx-G) protein to the PRE.[115] Perhaps more remarkably, they showed that the ncRNAs were required continually to control *Ubx* expression and could in fact be provided in *trans*. The amazing findings of the Sauer lab not only fit well with the data suggesting that transcription through a PRE would inactivate its silencing activity, but extending the role of ncRNAs as factors providing specificity to chromatin modifying enzymes.

However, just a few months later, the group of Alexander Mazo presented data on these same ncRNAs that directly contradicted Sauer's findings. According to Sauer's model, ncRNAs are continually required to activate *Ubx* expression. Mazo's group, using high-resolution, multiplex RNA in situ hybridization, showed that this cannot be the case because, although the ncRNAs are expressed in the *Ubx* domain of expression, they are, in fact, expressed in the cells not expressing *Ubx*.[116] Mazo also confirms older findings showing that the ncRNAs from the *bxd* region are only transiently expressed in the very early embryo and, therefore, cannot maintain *Ubx* expression. Later in 2007, the group of Welcome Bender reported new data definitively contradicting Sauer's findings. Using gene conversion Sipos et al (2007)[117] created specific deletions that removed the templates of these *bxd* ncRNA. These deletions resulted in neither loss of *Ubx* expression nor loss-of-function phenotypes indicating that the 3-kb region including the PRE is required for repression, but not for activation, of *Ubx*.

It is obvious from these three papers that the role of the ncRNAs in the BX-C is a complex problem. Confusions can arise between data sets due to a number of issues including the number of ncRNAs and their spatial and temporal expression profiles. Although the community is shifting away from the Sauer model, the work from the Mazo group is far from completely shutting the door on the idea that ncRNAs function to inactivate PREs. For example, Mazo only looks at one of the ncRNAs and the mechanism that he proposes to describe the effects that he observed cannot readily be translated to most of the other ncRNAs. Obviously, more work will be required to discover the consequences of the removal of these ncRNAs.

MicroRNAs in the BX-C

With the frenzied activity in the field of microRNAs, it is perhaps unsurprising that microRNAs have been found in ncRNAs of the BX-C. Thus far, the only microRNAs discovered in the BX-C lie in the *iab-3* domain, within one of the two regions of the BX-C transcribed on both strands. In fact, these miRNAs are the result of the same DNA hairpin being transcribed from both strands.[111,118-120] The antisense transcript produces miR-iab-4-5p and miR-iab-4-3p, while the sense transcript produces the miR-iab8-5p and miR-iab-8-3p. Although both sets of miRNAs are very similar in sequence (given that they are sense and antisense copies of a DNA hairpin), they are expressed in very different patterns. The antisense transcript, produced from the *iab4* ncRNA,[e] is expressed in the early embryo from PS8 to PS12. Its parasegment-specific expression persists throughout embryogenesis, but by late embryogenesis it is primarily restricted to the CNS. The sense transcript, on the other hand, seems to derive from the 120 kb ncRNA starting in the *iab-8 cis*-regulatory domain (see above). This transcript becomes visible starting at full germband extension (after the *iab-4* transcript) and is restricted to PS13 and 14. Like the *iab-4* transcript, this transcript persists throughout embryogenesis and becomes restricted to the CNS (though in PS13 and 14). Among the predicted targets for these microRNAs are the BX-C homeotic genes, *Ubx*, *abd-A* and *Abd-B*. Much work has gone into proving that these genes are the true targets of these microRNAs through overexpression. Although some repression of *Ubx* was found, these results had to be interpreted cautiously due to numerous caveats. For example, in PS8 to PS12 where iab-4-5-p is expressed, it is expressed in the same cells as the Abd-A protein, a protein known to repress *Ubx*.[21,121] Moreover, the *Ubx* expression pattern in PS7 where miR-iab-4-5p is not expressed is very similar to *Ubx* expression in PS8 where the microRNA is present. Therefore, it seems that if *Ubx* is the target of these miRNAs, their effects on *Ubx* expression are subtle.

Recently, Welcome Bender (2008)[111] described the mutant phenotypes of flies deleted for these miRNAs. As expected the phenotypes were quite subtle. Flies mutant for the miRNAs display no visible phenotypes, but are sterile. The exact cause of the sterility has not been completely elucidated, but at least in the males, it seems to be due to an inability to properly bend their abdomen to copulate. Interestingly, Bender also showed that the miRNAs responsible for these phenotypes come from the *iab-8* ncRNA as opposed to the *iab-4* transcript. As mentioned above, the *iab-8* transcript is expressed in PS13 and 14, where *Ubx* is normally not expressed. Accordingly, in the miRNA mutants, *Ubx* becomes slightly expressed in PS13. As the miRNAs may have more targets, is still not completely clear if this *Ubx* misexpression causes the sterility phenotype.

Conclusion

There is a cyclic nature to science. Old ideas often fall into disfavor, only to re-emerge in an evolved form decades later. The ideas surrounding the workings of the BX-C are no exception. The idea that the *iab* domains represented genes, now show some truth with the discovery of the ncRNAs. Even Ed Lewis's sequential reaction model has been revisited by the finding of Sauer and how the ncRNAs might target enzymes to the *cis*-regulatory domains. This cyclic nature just represents our ignorance in the complexity of biology.

During the almost 90 years of research on the BX-C, we have learned a lot, but we still cannot say that we understand how the BX-C works. Using both genetic and transgenic approaches, we have primarily been able to identify important regulatory elements and roughly map out the possible biological activities of these elements. Yet, even with all of this, we know little about how these elements work together to control BX-C homeotic gene expression. It is clear now that BX-C research is jumping to a new level of complexity. Hopefully, the new tools available in *Drosophila*, will finally allow us to answer these questions and allow for the next 90 years of BX-C research to be as fruitful as the first have been.

[e]When the ncRNA called the iab-4 RNA was discovered it was found in a region of DNA thought to be part of the *iab-4 cis*-regulatory domain. We now know from the precise mapping of the domains with enhancer trap lines that this region is actually part of *iab-3*. This is consistent with the pattern of expression of this iab-4 RNA beginning in PS8. To avoid confusion in the literature, the iab-4 nomenclature has been preserved.

References

1. Bridges CB, Morgan TH. The Third-Chromosome Group of Mutant Characters of Drosophila melanogaster. The Carnegie Institution, Washington, DC. 1923.
2. Garcia-Bellido A. Genetic control of wing disc development in Drosophila. Ciba Found Symp 1975; 0(29):161-182.
3. Hollander WF. Bithorax alleles. Dros Inf Serv 1937; 8:77.
4. Duncan I, Montgomery GEB. Lewis and the bithorax complex: part II. From cis-trans test to the genetic control of development. Genetics 2002; 161(1):1-10.
5. Duncan I, Montgomery GEB. Lewis and the bithorax complex: part I. Genetics 2002; 160(4):1265-1272.
6. Lewis EB. A gene complex controlling segmentation in Drosophila. Nature 1978; 276(5688):565-570.
7. Bender W, Spierer P, Hogness DS. Chromosomal walking and jumping to isolate DNA from the Ace and rosy loci and the bithorax complex in Drosophila melanogaster. J Mol Biol 1983; 168(1):17-33.
8. Bender W, Akam M, Karch F et al. Molecular genetics of the bithorax complex in Drosophila melanogaster. Science 1983; 221:23-29.
9. Karch F, Weiffenbach B, Peifer M et al. The abdominal region of the bithorax complex. Cell 1985; 43(1):81-96.
10. Sanchez-Herrero E, Vernos I, Marco R et al. Genetic organization of Drosophila bithorax complex. Nature 1985; 313(5998):108-113.
11. Tiong S, Bone LM, Whittle JR. Recessive lethal mutations within the bithorax-complex in Drosophila. Mol Gen Genet 1985; 200(2):335-342.
12. Casanova J, Sanchez-Herrero E, Busturia A et al. Double and triple mutant combinations of the bithorax complex of Drosophila. EMBO J 1987; 6:3103-3109.
13. Hogness DS, Lipshitz HD, Beachy PA et al. Regulation and products of the Ubx domain of the bithorax complex. Cold Spring Harb Symp Quant Biol 1985; 50:181-194.
14. Garber RL, Kuroiwa A, Gehring WJ. Genomic and cDNA clones of the homeotic locus Antennapedia in Drosophila. EMBO J 1983; 2(11):2027-2036.
15. Scott MP, Weiner AJ, Hazelrigg TI et al. The molecular organization of the Antennapedia locus of Drosophila. Cell 1983; 35(3 Pt 2):763-776.
16. McGinnis W, Levine MS, Hafen E et al. A conserved DNA sequence in homoeotic genes of the Drosophila Antennapedia and bithorax complexes. Nature 1984; 308(5958):428-433.
17. Scott MP, Weiner AJ. Structural relationships among genes that control development: sequence homology between the Antennapedia, Ultrabithorax and fushi tarazu loci of Drosophila. Proc Natl Acad Sci USA 1984; 81(13):4115-4119.
18. Regulski M, Harding K, Kostriken R et al. Homeo box genes of the Antennapedia and bithorax complexes of Drosophila. Cell 1985; 43(1):71-80.
19. Beachy PA, Helfand SL, Hogness DS. Segmental distribution of bithorax complex proteins during Drosophila development. Nature 1985; 313(6003):545-551.
20. Celniker SE, Sharma S, Keelan DJ et al. The molecular genetics of the bithorax complex of Drosophila: cis- regulation in the Abdominal-B domain. EMBO J 1990; 9(13):4277-4286.
21. Karch F, Bender W, Weiffenbach B. abdA expression in Drosophila embryos. Genes Dev 1990; 4:1573-1587.
22. Sanchez-Herrero E. Control of the expression of the bithorax complex genes abdominal-A and abdominal-B by cis-regulatory regions in Drosophila embryos. Development 1991; 111(2):437-449.
23. White RAH, Wilcox M. Regulation of the distribution of Ultrabithorax proteins in Drosophila. Nature 1985; 318(563-567).
24. Harding K, Wedeen C, McGinnis W et al. Spatially regulated expression of homeotic genes in Drosophila. Science 1985; 229(4719):1236-1242.
25. Martinez-Arias A, Lawrence P. Parasegments and compartments in the Drosophila embryo. Nature 1985; 313:639-642.
26. Peifer M, Karch F, Bender W. The bithorax complex: control of segmental identity. Genes Dev 1987; 1:891-898.
27. Little JW, Byrd CA, Brower DL. Effect of abx, bx and pbx mutations on expression of homeotic genes in Drosophila larvae. Genetics 1990; 124(4):899-908.
28. Cabrera C, Botas J, Garcia-Bellido A. Distribution of Ultrabithorax proteins in mutants of Drosophila bithorax complex and its trans-regulatory genes. Nature 1985; 318:569-571.
29. White RA, Akam M. Contrabithorax mutations cause inappropriate expression of Ultrabithorax products in Drosophila. Nature 1985; 318:567-569.
30. Irvine KD, Helfand SL, Hogness DS. The large upstream control region of the Drosophila homeotic gene Ultrabithorax. Development 1991; 111(2):407-424.

31. Lin Q, Chen Q, Lin L et al. Promoter targeting sequence mediates enhancer interference in the Drosophila embryo. Proc Natl Acad Sci USA 2007; 104(9):3237-3242.
32. Ingham PW. The molecular genetics of embryonic pattern formation in Drosophila. Nature 1988; 335(6185):25-34.
33. Hoch M, Jackle H. Transcriptional regulation and spatial patterning in Drosophila. Curr Opin Genet Dev 1993; 3(4):566-573.
34. Kornberg TB, Tabata T. Segmentation of the Drosophila embryo. Curr Opin Genet Dev 1993; 3(4):585-594.
35. DiNardo S, Heemskerk J, Dougan S et al. The making of a maggot: patterning the Drosophila embryonic epidermis. Curr Opin Genet Dev 1994; 4(4):529-534.
36. White RA, Lehmann R. A gap gene, hunchback, regulates the spatial expression of Ultrabithorax. Cell 1986; 47(2):311-321.
37. Irish VF, Martinez-Arias A, Akam M. Spatial regulation of the Antennapedia and Ultrabithorax homeotic genes during Drosophila early development. EMBO J 1989; 8(5):1527-1537.
38. Shimell MJ, Peterson AJ, Burr J et al. Functional analysis of repressor binding sites in the iab-2 regulatory region of the abdominal-A homeotic gene. Dev Biol 2000; 218(1):38-52.
39. Casares F, Sanchez Herrero E. Regulation of the infraabdominal regions of the bithorax complex of Drosophila by gap genes. Development 1995; 121(6):1855-1866 Issn: 0950-1991.
40. Duncan IM. Polycomblike: a gene that appears to be required for the normal expression of the bithorax and antennapedia gene complexes of Drosophila melanogaster. Genetics 1982; 102(1):49-70.
41. Jürgens G. A group of genes controlling the spatial expression of the bithorax complex in Drosophila. Nature 1985; 316(316):153-155.
42. Kennison JA, Tamkun JW. Dosage-dependent modifiers of polycomb and antennapedia mutations in Drosophila. Proc Natl Acad Sci USA 1988; 85(21):8136-8140.
43. Shearn A. The ash-1, ash-2 and trithorax genes of Drosophila melanogaster are functionally related. Genetics 1989; 121(3):517-525.
44. Paro R. Imprinting a determined state into the chromatin of Drosophila. Trends Genet 1990; 6(12):416-421 Issn: 0168-9525.
45. Schwartz YB, Pirrotta V. Polycomb silencing mechanisms and the management of genomic programmes. Nat Rev Genet 2007; 8(1):9-22.
46. Simon J, Peifer M, Bender W et al. Regulatory elements of the bithorax complex that control expression along the anterior-posterior axis. EMBO J 1990; 9(12):3945-3956 Issn: 0261-4189.
47. Qian S, Capovilla M, Pirrotta V. The bx region enhancer, a distant cis-control element of the Drosophila Ubx gene and its regulation by hunchback and other segmentation genes. EMBO J 1991; 10(6):1415-1425.
48. Muller J, Bienz M. Sharp anterior boundary of homeotic gene expression conferred by the fushi tarazu protein. EMBO J 1992; 11(10):3653-3661 Issn: 0261-4189.
49. Zhou J, Ashe H, Burks C et al. Characterization of the transvection mediating region of the abdominal- B locus in Drosophila. Development 1999; 126(14):3057-3065.
50. Barges S, Mihaly J, Galloni M et al. The Fab-8 boundary defines the distal limit of the bithorax complex iab-7 domain and insulates iab-7 from initiation elements and a PRE in the adjacent iab-8 domain. Development 2000; 127:779-790.
51. Shimell MJ, Simon J, Bender W et al. Enhancer point mutation results in a homeotic transformation in Drosophila. Science 1994; 264(5161):968-971 Issn: 0036-8075.
52. Brock HW, van Lohuizen M. The Polycomb group—no longer an exclusive club? Curr Opin Genet Dev 2001; 11(2):175-181.
53. Muller J, Bienz M. Long range repression conferring boundaries of Ultrabithorax expression in the Drosophila embryo. EMBO J 1991; 10(11):3147-3155 Issn: 0261-4189.
54. Simon J, Chiang A, Bender W et al. Elements of the Drosophila bithorax complex that mediate repression by Polycomb group products. Dev Biol 1993; 158(1):131-144 Issn: 0012-1606.
55. Chan CS, Rastelli L, Pirrotta V. A Polycomb response element in the Ubx gene that determines an epigenetically inherited state of repression. EMBO J 1994; 13(11):2553-2564 Issn: 0261-4189.
56. Fritsch C, Brown JL, Kassis JA et al. The DNA-binding Polycomb group protein Pleiohomeotic mediates silencing of a Drosophila homeotic gene. Development 1999; 126(17):3905-3913.
57. Busturia A, Lloyd A, Bejarano F et al. The MCP silencer of the Drosophila Abd-B gene requires both Pleiohomeotic and GAGA factor for the maintenance of repression. Development 2001; 128(11):2163-2173.
58. Chiang A, O'Connor MB, Paro R et al. Discrete Polycomb-binding sites in each parasegmental domain of the bithorax complex. Development 1995; 121(6):1681-1689 Issn: 0950-1991.
59. Kassis JA. Unusual properties of regulatory DNA from the Drosophila engrailed gene: three "pairing-sensitive" sites within a 1.6-kb region. Genetics 1994; 136(3):1025-1038.

60. Gindhart JG, Jr., Kaufman TC. Identification of Polycomb and trithorax group responsive elements in the regulatory region of the Drosophila homeotic gene Sex combs reduced. Genetics 1995; 139(2):797-814 Issn: 0016-6731.
61. Hagstrom K, Muller M, Schedl P. A Polycomb and GAGA dependent silencer adjoins the Fab-7 boundary in the Drosophila bithorax comple. Genetics 1997; 146:1365-1380.
62. Busturia A, Bienz M. Silencers in abdominal-B, a homeotic Drosophila gene. EMBO J 1993; 12(4):1415-1425 Issn: 0261-4189.
63. Pirrotta V, Chan CS, McCabe D, Qian S. Distinct parasegmental and imaginal enhancers and the establishment of the expression pattern of the Ubx gene. Genetics 1995; 141(4):1439-1450.
64. Mihaly J, Barges S, Sipos L et al. Dissecting the regulatory landscape of the Abd-B gene of the bithorax complex. Development 2006; 133(15):2983-2993.
65. O'Kane CJ, Gehring WJ. Detection in situ of genomic regulatory elements in Drosophila. Proc Natl Acad Sci USA 1987; 84(24):9123-9127.
66. Bender W, Hudson A. P element homing to the Drosophila bithorax complex. Development 2000; 127(18):3981-3992.
67. Gyurkovics H, Gausz J, Kummer J et al. A new homeotic mutation in the Drosophila bithorax complex removes a boundary separating two domains of regulation. EMBO J 1990; 9(8):2579-2585 Issn: 0261-4189.
68. Karch F, Galloni M, Sipos L et al. Mcp and Fab-7: molecular analysis of putative boundaries of cis-regulatory domains in the bithorax complex of Drosophila melanogaster. Nucleic Acids Res 1994; 22(15):3138-3146 Issn: 0305-1048.
69. Mihaly J, Hogga I, Gausz J et al. In situ dissection of the Fab-7 region of the bithorax complex into a chromatin domain boundary and a Polycomb-response element. Development 1997; 124(9):1809-1820.
70. Mihaly J, Hogga I, Barges S et al. Chromatin domain boundaries in the Bithorax complex. Cell Mol Life Sci 1998; 54(1):60-70.
71. Galloni M, Gyurkovics H, Schedl P et al. The bluetail transposon: evidence for independent cis-regulatory domains and domain boundaries in the bithorax complex. EMBO J 1993; 12(3):1087-1097 Issn: 0261-4189.
72. Vazquez J, Farkas G, Gaszner M et al. Genetic and Molecular Analysis of Chromatin Domains. Cold Spring Harb Symp Quant Biol 1993; 58:45-54.
73. Lewis EB. Pseudoallelism and gene evolution. Cold Spring Harb Symp Quant Biol 1951; 16:159-174.
74. Lewis EB. The theory and application of a new method of detecting chromosomal rearrangements in Drosophila melanogaster. Am Nat 1954; 88:225-239.
75. Gans M. Etude génétique et physiologique du mutant z de Drosophila melanogaster. Bull Biol Fr Belg 1953;(Suppl 38):1-90.
76. Morris JR, Geyer PK, Wu CT. Core promoter elements can regulate transcription on a separate chromosome in trans. Genes Dev 1999; 13(3):253-258.
77. Morris JR, Chen J, Filandrinos ST et al. An analysis of transvection at the yellow locus of Drosophila melanogaster. Genetics 1999; 151(2):633-651.
78. Martinez-Laborda A, Gonzalez-Reyes A, Morata G. Trans regulation in the Ultrabithorax gene of Drosophila: alterations in the promoter enhance transvection. EMBO J 1992; 11(10):3645-3652.
79. Hopmann R, Duncan D, Duncan I. Transvection in the iab-5,6,7 region of the bithorax complex of Drosophila: homology independent interactions in trans. Genetics 1995; 139(2):815-833 Issn: 0016-6731.
80. Hendrickson JE, Sakonju S. Cis and trans interactions between the iab regulatory regions and abdominal-A and abdominal-B in Drosophila melanogaster. Genetics 1995; 139(2):835-848 Issn: 0016-6731.
81. Sipos L, Mihaly J, Karch F et al. Transvection in the drosophila abd-B domain. Extensive upstream sequences are involved in anchoring distant cis-regulatory regions to the promoter. Genetics 1998; 149(2):1031-1050.
82. Kassis JA, VanSickle EP, Sensabaugh SM. A fragment of engrailed regulatory DNA can mediate transvection of the white gene in Drosophila. Genetics 1991; 128(4):751-761 Issn: 0016-6731.
83. Muller M, Hagstrom K, Gyurkovics H et al. The mcp element from the Drosophila melanogaster bithorax complex mediates long-distance regulatory interactions. Genetics 1999; 153(3):1333-1356.
84. Pirrotta V, Rastelli L. White gene expression, repressive chromatin domains and homeotic gene regulation in Drosophila. Bioessays 1994; 16(8):549-556 Issn: 0265-9247.
85. Sigrist CJ, Pirrotta V. Chromatin insulator elements block the silencing of a target gene by the Drosophila polycomb response element (PRE) but allow trans interactions between PREs on different chromosomes. Genetics 1997; 147(1):209-221.

86. Vazquez J, Muller M, Pirrotta V et al. The Mcp element mediates stable long-range chromosome-chromosome interactions in Drosophila. Mol Biol Cell 2006; 17(5):2158-2165.
87. Bantignies F, Grimaud C, Lavrov S et al. Inheritance of Polycomb-dependent chromosomal interactions in Drosophila. Genes Dev 2003; 17(19):2406-2420.
88. Fedorova E, Sadoni N, Dahlsveen IK et al. The nuclear organization of Polycomb/Trithorax group response elements in larval tissues of Drosophila melanogaster. Chromosome Res 2008; 16(4):649-673.
89. Saurin AJ, Shiels C, Williamson J et al. The human polycomb group complex associates with pericentromeric heterochromatin to form a novel nuclear domain. J Cell Biol 1998; 142(4):887-898.
90. Buchenau P, Hodgson J, Strutt H et al. The distribution of polycomb-group proteins during cell division and development in Drosophila embryos: impact on models for silencing. J Cell Biol 1998; 141(2):469-481.
91. Fauvarque MO, Dura JM. polyhomeotic regulatory sequences induce developmental regulator- dependent variegation and targeted P-element insertions in Drosophila. Genes Dev 1993; 7(8):1508-1520.
92. Hama C, Ali Z, Kornberg TB. Region-specific recombination and expression are directed by portions of the Drosophila engrailed promoter. Genes Dev 1990; 4(7):1079-1093.
93. Kassis JA, Noll E, VanSickle EP et al. Altering the insertional specificity of a Drosophila transposable element. Proc Natl Acad Sci USA 1992; 89(5):1919-1923 Issn: 0027-8424.
94. Cleard F, Moshkin Y, Karch F et al. Probing long-distance regulatory interactions in the Drosophila melanogaster bithorax complex using Dam identification. Nat Genet 2006; 38(8):931-935.
95. Maeda RK, Karch F. Making connections: boundaries and insulators in Drosophila. Curr Opin Genet Dev 2007; 17(5):394-399.
96. Valenzuela L, Kamakaka RT. Chromatin insulators. Annu Rev Genet 2006; 40:107-138.
97. Kellum R, Schedl P. A position-effect assay for boundaries of higher order chromosomal domains. Cell 1991; 64(5):941-950.
98. Kellum R, Schedl P. A group of scs elements function as domain boundaries in an enhancer- blocking assay. Mol-Cell-Biol 1992; 12:2424-2431.
99. Hagstrom K, Muller M, Schedl P. Fab-7 functions as a chromatin domain boundary to ensure proper segment specification by the Drosophila bithorax complex. Genes Dev 1996; 10:3202-3215.
100. Zhou J, Barolo S, Szymanski P et al. The Fab-7 element of the bithorax complex attenuates enhancer-promoter interactions in the Drosophila embryo. Genes Dev 1996; 10(24):3195-3201.
101. Gruzdeva N, Kyrchanova O, Parshikov A et al. The Mcp Element from the bithorax Complex Contains an Insulator That Is Capable of Pairwise Interactions and Can Facilitate Enhancer-Promoter Communication. Mol Cell Biol 2005; 25(9):3682-3689.
102. Zhou J, Levine M. A novel cis-regulatory element, the PTS, mediates an anti-insulator activity in the Drosophila embryo. Cell 1999; 99(6):567-575.
103. Chen Q, Lin L, Smith S et al. Multiple Promoter Targeting Sequences exist in Abdominal-B to regulate long-range gene activation. Dev Biol 2005; 286:629-636.
104. Lin Q, Wu D, Zhou J. The promoter targeting sequence facilitates and restricts a distant enhancer to a single promoter in the Drosophila embryo. Development 2003; 130(3):519-526.
105. Akbari OS, Bae E, Johnsen H et al. A novel promoter-tethering element regulates enhancer-driven gene expression at the bithorax complex in the Drosophila embryo. Development 2008; 135:123-131.
106. Lipshitz HD, Peattie DA, Hogness DS. Novel transcripts from the Ultrabithorax domain of the bithorax complex. Genes Dev 1987; 1(3):307-322.
107. Sanchez-Herrero E, Akam M. Spatially ordered transcription of regulatory DNA in the bithorax complex of Drosophila. Development 1989; 107(2):321-329.
108. Cumberledge S, Zaratzian A, Sakonju S. Characterization of two RNAs transcribed from the cis-regulatory region of the abd-A domain within the Drosophila bithorax complex. Proc Natl Acad Sci USA 1990; 87(9):3259-3263.
109. Bae E, Calhoun VC, Levine M et al. Characterization of the intergenic RNA profile at abdominal-A and Abdominal-B in the Drosophila bithorax complex. Proc Natl Acad Sci USA 2002; 99(26):16847-16852.
110. Drewell RA, Bae E, Burr J et al. Transcription defines the embryonic domains of cis-regulatory activity at the Drosophila bithorax complex. Proc Natl Acad Sci USA 2002; 99(26):16853-16858.
111. Bender W. MicroRNAs in the Drosophila bithorax complex. Genes Dev 2008; 22(1):14-19.
112. Hogga I, Karch F. Transcription through the iab-7 cis-regulatory domain of the bithorax complex interferes with maintenance of Polycomb-mediated silencing. Development 2002; 129(21):4915-4922.
113. Bender W, Fitzgerald DP. Transcription activates repressed domains in the Drosophila bithorax complex. Development 2002; 129(21):4923-4930.

114. Rank G, Prestel M, Paro R. Transcription through intergenic chromosomal memory elements of the Drosophila bithorax complex correlates with an epigenetic switch. Mol Cell Biol 2002; 22(22):8026-8034.
115. Sanchez-Elsner T, Gou D, Kremmer E et al. Noncoding RNAs of trithorax response elements recruit Drosophila Ash1 to Ultrabithorax. Science 2006; 311(5764):1118-1123.
116. Petruk S, Sedkov Y, Riley KM et al. Transcription of bxd noncoding RNAs promoted by trithorax represses Ubx in cis by transcriptional interference. Cell 2006; 127(6):1209-1221.
117. Sipos L, Kozma G, Molnar E et al. In situ dissection of a Polycomb response element in Drosophila melanogaster. Proc Natl Acad Sci USA 2007; 104(30):12416-12421.
118. Tyler DM, Okamura K, Chung WJ et al. Functionally distinct regulatory RNAs generated by bidirectional transcription and processing of microRNA loci. Genes Dev 2008; 22(1):26-36.
119. Ronshaugen M, Biemar F, Piel J et al. The Drosophila microRNA iab-4 causes a dominant homeotic transformation of halteres to wings. Genes Dev 2005; 19(24):2947-2952.
120. Stark A, Bushati N, Jan CH et al. A single Hox locus in Drosophila produces functional microRNAs from opposite DNA strands. Genes Dev 2008; 22(1):8-13.
121. Macias A, Casanova J, Morata G. Expression and regulation of the abd-A gene of Drosophila. Development 1990; 110(4):1197-1207.
122. Benzer S. Fine structure of a genetic region in bacteriophage. Proc Natl Acad Sci USA 1955; 41:344-354.
123. Martin CH, Mayeda CA, Davis CA et al. Complete sequence of the bithorax complex of Drosophila. Proc Natl Acad Sci USA 1995; 92(18):8398-8402 Issn: 0027-8424.

CHAPTER 3

Maintenance of Hox Gene Expression Patterns

Samantha Beck, Floria Faradji, Hugh Brock and Frédérique Peronnet*

Abstract

Once established, homeotic gene (Hox) expression is maintained in the original pattern by Polycomb-group (PcG) and trithorax-group (trxG) proteins therefore named maintenance proteins (MPs). PcG and trxG proteins maintain silencing and activation of Hox and many other genes, respectively. We provide here a brief overview of genetics and molecular biology of these proteins and of a third class of proteins termed Enhancers of Trithorax and Polycomb (ETP) that are required for both maintenance of silencing and activation of Hox genes. We examine the recruitment of MPs onto maintenance elements (MEs), their role in the regulation of transcription and the epigenetic marks that could provide maintenance. Lastly, we discuss two important roles of PcG proteins in replication of DNA and stem cell renewal and maintenance.

Introduction

Changes in gene expression patterns define cell identity in development. Specific gene expression patterns must be passed on to daughter cells during cell division, in a process termed transcriptional maintenance.[1] Disruption of maintenance can have disastrous effects in development and disease.[1] Maintenance must be epigenetic because the same DNA sequence is inherited by all cells, yet the gene expression patterns differ.[2,3] Epigenetic regulation is widespread in eukaryotes and includes mating type silencing in yeast, position-effect variegation, gametic imprinting and dosage compensation. Epigenetic regulation in different systems shares mechanisms including DNA methylation, histone modifications, use of histone variants, nucleosome remodeling, nuclear compartmentalization and specific higher order chromatin structures.[4] One of the key challenges in development is to understand the mechanisms of maintenance in detail.

The proteins required for maintenance of gene expression patterns are "maintenance proteins" (MPs).[5] The best-characterized MPs are encoded by the Polycomb group (PcG) and trithorax group (trxG) genes, which maintain repression (silencing) and activation respectively of Hox and other genes. About 20 PcG and 15 trxG genes have been characterized in Drosophila.[5] Most PcG and trxG proteins have homologs with similar functions in other eukaryotes.[5] MPs regulate many developmental processes including stem cell self-renewal, X-chromosome inactivation, the hedgehog signaling pathway, genomic imprinting, senescence, the cell cycle and DNA replication.[1,5,6] The structure and function of Drosophila and mammalian Hox complexes is covered in other chapters of this volume and we refer readers there for background not covered here. In this review we will provide a brief overview of genetics and molecular biology of PcG and trxG proteins, discuss what is known about their regulatory elements in Hox clusters and point out some

*Corresponding Author: Frédérique Peronnet—UMR 7622—Biologie du Développement; CNRS—Université Pierre et Marie Curie Paris 6 (UPMC), 9, quai Saint-Bernard, 75005 Paris, France. Email: frederique.peronnet@upmc.fr

Hox Genes: Studies from the 20th to the 21st Century, edited by Jean S. Deutsch.

unanswered questions about the role of MPs and their regulatory elements in Hox gene regulation. We will also discuss genes whose products are required for both maintenance of transcription and for silencing of Hox genes. For genetic reasons these genes have been termed "Enhancers of Trithorax and Polycomb" or ETPs.[7]

Genetics of PcG and trxG Genes

PcG mutations were discovered more than 60 years ago because they had homeotic phenotypes, but did not map to what became known as the Hox gene complexes. Ed Lewis, in his seminal review of the Bithorax Complex (BX-C),[8] pointed out that the phenotypes of PcG mutants in the posterior thorax and abdomen were posterior transformations, consistent with derepression of BX-C coding genes and concluded that the function of PcG genes was to repress Hox genes. This idea was soon confirmed molecularly. In PcG mutations, *Ultrabithorax* (*Ubx*) expression is de-repressed in the anterior parasegments of the embryos.[9]

PcG mutants themselves have homeotic transformations.[10] Most of these are posterior transformations, but confusingly, the extra sex combs that were the key phenotype of the first PcG gene to be discovered (*extra sex combs* (*esc*))[11] is an example of an anterior transformation, as the legs arising from the second and third thoracic segments are transformed to resemble the leg from the first thoracic segment. This phenotype arises from derepression of *Sex combs reduced* (*Scr*) in the second and third thoracic segments.[12] About a dozen PcG mutations have been tested and all cause derepression of Hox genes in embryos.[13]

Another key identifying characteristic of PcG mutations is that double mutants of different PcG mutations have stronger phenotypes than either single mutant, which led Jürgens to propose the important idea that PcG proteins have similar functions or act in parallel pathways.[14] Genetic screens for enhancers of PcG phenotypes produced estimates of 30-40 PcG genes.[14,15] However, some PcG mutants, like *Enhancer of Polycomb* (*E*(*Pc*))[16] and *Suppressor of zeste 2* (*Su*(*z*)*2*)[17] do not have homeotic phenotypes and/or do not cause derepression of Hox genes. *E*(*Pc*) encodes a component of a histone acetyl-transferase complex associated with transcriptional activation and thus likely acts indirectly to promote transcription of other PcG genes. More recently it has been suggested that enhancers of PcG phenotypes that do not themselves have homeotic phenotypes should be excluded from potential membership in the PcG unless it can be demonstrated that they have a direct role in PcG-mediated silencing.[18] A screen for new PcG genes in Drosophila based on possession of homeotic phenotypes in discs suggests that most PcG genes have already been discovered and that the number is unlikely to exceed 20.[18]

The first trxG mutation, *trithorax*, (*trx*) was discovered on the basis of its homeotic transformation.[19] However, the *trx* phenotype is opposite to that of PcG mutations, as mutants exhibit anterior transformations in the posterior thorax and abdomen. This phenotype is consistent with a role for trxG genes in activation of Hox genes. As expected, *trx* mutant embryos show decreased expression of *Ubx* in posterior parasegments.[20]

Ingham first reported the interesting observation that *trx* mutations suppressed the posterior transformations of *esc*.[21] If PcG phenotypes arise from derepression of Hox genes and trxG phenotypes arise from decreased transcription of Hox genes, then in double mutants the homeotic genes cannot be strongly derepressed and therefore the PcG phenotype should be prevented. The property that trxG mutations suppress PcG phenotypes was exploited to screen for other trxG mutations.[22] About 15 trxG genes have been described so far.[23] There are many classes of molecules required for transcriptional activation and potentially trxG genes should be more numerous and have more diverse functions than PcG genes.[23]

In a screen for deficiencies that enhance or suppress the third to second leg transformation of some trxG mutant alleles, a homeotic phenotype due to loss of *Ubx* and gain of *Antennapedia* (*Antp*) expression in the third leg, Gildea et al[7] identified 5 PcG genes {*Enhancer of zeste* (*E*(*z*)), *E*(*Pc*), *Su*(*z*)2, *Posterior sex combs* (*Psc*), *Additional sex combs* (*Asx*) and *Sex comb on midleg* (*Scm*)} whose mutations unexpectedly enhance the phenotype of the trxG *absent, small, or homeotic discs 1* (*ash1*), *brahma* (*brm*) and *trx* mutants. They proposed that these genes be called Enhancers of trihorax and Polycomb (ETPs). The phenotypes of ETP mutants imply that they function in maintenance of both activation and silencing.

In Drosophila, mutants of the ETP gene *Asx* exhibit simultaneous anterior and posterior transformations and enhance homeotic phenotypes of both PcG and trxG mutations.[24] Interestingly, mutant embryos for mouse *Asxl1*, one of the three *Asx* homologs, also exhibit simultaneous anterior and posterior homeotic transformations of the antero-posterior axis consistent with Hox gene misregulation[5]. This suggests that the function of ETP is conserved throughout evolution.

Other ETP genes exist in Drosophila. Mutants of the *Trithorax-like* (*Trl*) gene present trxG phenotypes, enhance the phenotype of *Ubx* loss of function alleles but also enhance the PcG phenotype of *Pc* alleles.[7,25] Hence, *Trl* also behaves both like a trxG and a PcG gene. *corto* also behaves genetically as an ETP as its mutants exhibit PcG as well as trxG phenotypes. Both loss of function and overexpression of *corto* promote the ectopic anterior expression of *Ubx*.[26] Furthermore, a loss-of-function allele of *corto* enhances the phenotype of mutants of the trxG gene *osa* as well as the PcG phenotype of the PcG genes *Polycomb* (*Pc*), *Polycomblike* (*Pcl*) and *polyhomeotic* (*ph*).[26,27] Lastly, a *dorsal switch protein 1* (*dsp1*) null allele enhances the haltere to wing transformation of several trxG mutants (*ash1*, *brm*, *osa*, *trx*), but male hemizygotes for this allele exhibit a PcG phenotype.[28] Furthermore, overexpression of *dsp1* results in enhancement of the polycomb phenotype. Thus, *dsp1* also genetically behaves like an ETP.

PcG Proteins and Their Complexes

The molecular biology of PcG proteins and their complexes has been extensively reviewed recently.[5,29,30] Here we briefly summarize our current understanding, but refer readers to earlier reviews for more detail on what is known about the identity and function of PcG proteins and to Table 1.

The first PcG complex purified was termed PcG Repressive Complex 1 (PRC1). PRC1 contains stoichiometric amounts of 4 PcG proteins: PC, PH, PSC and dRing and smaller amounts of SCM and many other proteins including Zeste, TATA-binding protein (TBP) and TBP-associated factors (TAFs), chaperones and actin homologs. In mammals PRC1 contain the same PcG subunits, but lacks most of the nonPcG subunits found in flies. The known biochemical functions of PRC1 can be reconstituted with just the PcG proteins.

The dRing subunit of PRC1 and the mammalian homologs Ring1A and Ring1B, are E3 ubiquitin ligases that ubiquitinate histone H2A on lysine 119 (H2AK119), a modification that is essential for silencing of Hox genes.[31] In mammalian PRC1 complexes, the PSC homolog Bmi-1 and Ring1a greatly stimulate the E3 ubiquitin ligase activity of Ring1b. In Drosophila, PRC1 also antagonizes the ATP-dependent nucleosome remodelling activity of SWI/SNF complexes[32] and can compact nucleosomal arrays.[33] These two activities can be carried out by PSC alone. The other subunits increase the overall activity in the assays, but their precise role is not fully understood. PRC1 could be recruited via binding of the chromodomain of PC to Histone 3 trimethylated on lysine 27 (H3K27me3).[34]

A second complex containing PcG proteins is named PRC2. In Drosophila, PRC2 contains ESC, E(Z), Su(Z)12 and the histone binding protein NURF55/CAF1.[35] In mammals, it contains the ESC homolog EED, E(Z), SUZ12 and the NURF55 homologs RbAp46/RbAp48. Variant PRC2 complexes have been isolated that differ by the presence of different EED isoforms and have been named PRC3 and PRC4.

The E(Z) subunit is a SET-domain histone methyltransferase that trimethylates H3K27, the histone mark associated with silencing.[36-38] However PRC2 is also required to mono- and dimethylate H3K27, two nearly ubiquitous histone modifications.[39] EZH2, the human homolog of E(Z), shows a preference for methylation of dinucleosomes over mononucleosomes which may be stimulated by linker histone H1 which interacts with PRC2 members.[40] The complex may have reduced specificity for H3K9, but the role of K9 methylation in PcG dependent silencing, if any, is not well understood. The function of this modification is unknown but is important for transcriptional repression by EZH2 and could recruit Heterochromatin Protein 1 (HP1).[41]

Table 1. Maintenance proteins and their function. See text for details and references.

Drosophila	Human	Domains	Activity	Complexes
PcG proteins				
Polycomb (PC)	HPC1,2, CBX6,7,8	Chromodomain	H3K27me3 binding	PRC1
Polyhomeotic (PH)	PHC1 (RAE28), PHC2,3	FCS Zn fingers, SAM	-	
Posterior Sex Comb (PSC)	BMI1	RING Zn fingers	-	
dRing	RING1, RNF2	RING Zn fingers	E3 ubiquitin ligase	
Enhancer of Zeste (E(Z)	EZH1, EZH2	SET, SANT	H3K27 methyltransferase	PRC2
Extra sex comb (ESC)	EED	WD40 repeats	-	
NURF55/CAF1	RbAp48/RBBP4	WD40 repeats	-	
Suppressor of zeste-12 (Su(Z)12)	SU(Z)12	Zn fingers	-	
Polycomb-like (Pcl)	HPCL1, HPCL2, HPCL 3	PHD Zn fingers, Tudor		Pcl-PRC2
Pho	YY1,2	Zn fingers	DNA binding	PhoRC
Pho-like (Pho-L)	-	Zn fingers	DNA binding	PhoRC
E(PC)	EPC1,2	-	-	-
dMi2	CHD3,4	Chromodomain, Zn fingers	-	-
TrxG proteins				
Trithorax (Trx)	MLL1,5	SET, Zinc fingers	H3K4 methyltransferase	TAC1
Absent small or homeotic discs 1 (Ash1)	ASH1L	SET, AT hook, Zn fingers, BAH	H3K4, H3K9, H4K20 methyltransferase	ASH1

continued on next page

Table 1. Continued

Drosophila	Human	Domains	Activity	Complexes
Absent small or homeotic discs 2 (Ash2)	ASH2L	PHD Zn fingers, SPRY	-	-
Brahma (BRM)	BRG1, hBRM	Bromodomain, BRK, helicase	AT Pase	BRM
OSA	hELD/OSA1	ARID	-	BRM
Moira (MOR)	SMARCC1,2	SANT, SWIRM	-	BRM
Kismet (KIS)	-	Chromodomain, helicase, SANT, BRK	DNA helicase	-
Kohtalo (KTO)	-	-	-	Mediator
Skuld	MED13,13L	TRAP240	-	Mediator
ETPs and/or cofactors				
Additional Sex Comb (ASX)	ASXL1,2,3	-	-	-
Corto	-	Chromodomain	-	-
Cramped (CRM)	CRAMP1L	SANT	-	-
DSP1	HMGB1	HMG	DNA binding	-
GAGA factor (GAF)	-	BTB/POZ, Zn fingers	DNA binding	-
Grainyhead	GRHL1,2,3	-	DNA binding	-
Lolal/Batman	ZFP161	BTB/POZ	DNA binding	-
Pipsqueak (PSQ)	-	BTB/POZ	DNA binding	-
Su(Z)2	-	Zn fingers	-	-
Zeste (Z)	-	Homeodomain-like	DNA binding	-

The ESC subunit increases the histone methyltransferase activity of E(Z) and the Su(Z)12 and NURF-55 subunits are essential for nucleosome binding.[42]

Other variants of PRC2 complexes have been identified. The Pcl-PRC2 complex contains the PcG protein PCL in addition to the 4 subunits found in PRC2 and the chaperone Heat-Shock Cognate 70-4 (HSC70-4), which also cofractionates with PRC1.[43] Importantly, loss of PCL prevents trimethylation of H3K27, suggesting that this protein has a key role in silencing-specific methylation.[43] This is consistent with observations that mono- and dimethylated H3K27 (H3K27me1, H3K27me2) are genome-wide[44] and that a key distinguishing feature of silenced genes is H3K27me3.[45]

A novel PcG complex was purified recently.[46] The Pho Repressive Complex (PhoRC) contains the only PcG protein with sequence-specific DNA binding ability, Pleiohomeotic (Pho)[47] and its homolog Pho-like (Pho-L).[48] It also contains a novel PcG protein, dSfmbt. dSfmbt contains a domain that recognizes mono- or dimethyl-lysine residues of histone H3K9 (H3K9me1, H3K9me2) and H4K20 (H4K20me1, H4K20me2), but not trimethylated H3K9 (H3K9me3) or H3K27 (H3K27me3). Dsfmbt also binds directly to Pho and Pho-L, but there is no evidence that PhoRC contacts PRC1 or PRC2 complexes.

Very recently another distinct PcG complex has been identified in Drosophila larvae, but apart from PCL, its subunit composition is not characterized.[49]

TrxG Proteins and Their Complexes

TrxG proteins are listed in Table 1. A complex containing Trx was first purified from Drosophila and named Trx Activating Complex 1 (TAC1).[50] Trx is a SET domain histone methyltransferase specific for H3K4. In addition, TAC1 contains the histone acetyltransferase CREB binding protein (CBP) and the SET binding factor 1 (SBF1). In mammals, most complexes containing the Trx homologs MLL1-4 resemble the COMPASS complex of yeast.[51] These complexes and the homologous yeast COMPASS complex all contain SET domain H3K4 methyltransferases, including SET1, or MLL1, MLL2, or MLL3,4. They also share 4 additional subunits: another trxG protein Absent Small Homeotic 2 (ASH2), the WD40 domain proteins RBBP5, WDR5 and a protein linked to dosage compensation, hDPY30. In addition, each complex has specific subunits. ASH1 is a member of a complex distinct from those containing Trx.[52] ASH1 is a histone methyltransferase whose specificity has been disputed as it has been reported to methylate H3K4, H3K9 and H4K20,[53] H3K4[54] and H3K36.[55] The Trx and ASH1 complexes might not function independently. Binding of Trx and ASH1 overlap nearly completely on polytene chromosomes and Trx binding is nearly completely abolished in *ash1* mutants.[56] In addition, Trx interacts directly with ASH1[57] and SNR1.[58]

Most of the molecularly characterized trxG proteins are members of ATP-dependent nucleosome remodelling complexes that slide or displace nucleosomes (reviewed in ref. 59). The SWI/SNF complex of yeast is called BAP in Drosophila and BAF in mammals. BAP contains the trxG proteins Moira, Osa, BRM, an ATP-ase, actin-related proteins, actin and SNR1, as well as other proteins. The mammalian complex is similar except that it contains BRG1 instead of BRM. A related complex (PBAP in flies or PBAF in mammals) contains the proteins above with the exception of Osa and contains the polybromodomain proteins (BAP170 or BAF180).[59] Other ATP-dependent nucleosome remodelling complexes could contribute to trxG function. The NURF complex is required to maintain Hox expression.

The trxG proteins Skuld and Kothalo are homologs of mammalian TRAP240 and TRAP230 and are subunits of the Drosophila mediator complex.[60]

ETP Proteins

ETPs are listed in Table 1. Except for E(Z), PSC and SCM, none of the ETPs have been found in PcG or in trxG core complexes, but molecular interactions between ETPs and these complexes have been demonstrated. For example, Corto co-immunoprecipitates with ESC and PC in embryonic extracts[61] and Su(Z)2 and GAF co-immunoprecipitate with PC,[62] suggesting that the ETPs can transiently interact with PcG and trxG complexes. In addition, Corto directly interacts with GAF and with the HMGB protein DSP1 suggesting that some ETPs are involved in collaborative processes.[63]

Recently, Grimaud et al[64] argued that molecular criteria should supersede phenotypic criteria for classification of maintenance proteins and proposed to keep the label PcG for members of silencing complexes and the label trxG for members of complexes that counteract PcG-mediated silencing. Thus, the term ETP should be kept for those maintenance proteins that play a dual role in PcG and trxG functions without belonging to any PcG or trxG complexes identified so far. GAF and the HMG protein DSP1 which recruites PcG complexes to Maintenance Elements (MEs)[62,65] (see below) fall into this category. The two ETPs Corto and DSP1 are simultaneously found on a *Scr* ME when active, whereas Corto alone is found on the same ME when inactive suggesting that different combinations of ETPs could favor the recruitment of either PcG or trxG complexes, thus participating in the maintenance of the silenced or active state of MEs.[63]

PcG and trxG Response Elements

PcG and trxG proteins bind complex DNA regulatory elements called PcG Response Elements (PREs) or trxG response elements (TREs). Regions containing PRE can contain TRE. Although intermingled, TREs are physically separable from PREs.[66] Therefore the regulatory sequences have also been called PRE/TRE. They are also called Cellular Memory Modules (CMM) or Maintenance Elements (ME) to reflect function in both maintenance of transcription and silencing (reviewed in ref. 67). In this review we will use PRE or TRE when referring to properties specific for maintenance of silencing or transcription respectively and ME for general properties of PREs and TREs.

Transgenes containing the promoter, stage- or tissue-specific Hox enhancers that governed expression of a reporter showed correct spatial regulation initially, but this regulation was not maintained for more that a few hours of development.[68] Therefore a search was initiated for sequences that allowed such transgenes to maintain spatial regulation and PREs were identified. In Drosophila, Hox PRE sequences share many properties with enhancers as they act at distance (up to 60 kb) and are position and orientation independent.[69] In mammals, a PRE has not been identified yet.

In general, transgenes containing PREs do not maintain silencing as effectively as the endogenous locus. At the *bithoraxoid* (*bxd*) PRE, longer transgenes silence more completely and later in development than shorter transgenes.[68] The regulatory regions of the BX-C are long, poorly characterized and complex, so one simple explanation is that the transgenes studied did not have all the necessary enhancers or ME sequences needed to reproduce the silencing achieved by the endogenous locus. Another possibility is that transgenes inserted randomly in the genome do not behave like the endogenous sequences because the chromatin domain at the site of insertion interferes with silencing. Moreover, as PREs and TREs can be intermingled in MEs, transgenes may carry sequences that could activate in some locations and silence in others, complicating interpretation. Finally, different MEs cooperate at their target loci to maintain stable responses. For instance, multiple PREs of the BX-C cluster in space in the nucleus,[70] possibly reinforcing silencing.

A very important recent study examined for the first time deletions at the endogenous *bxd* PRE.[71] Deletions that included a 665 bp region already identified as a PRE exhibited transformation of anterior segments towards posterior ones in adult flies, similar to those seen in PcG mutants and caused derepression of *Ubx* in embryos. To address the problem of potential redundancy within this ME, the authors deleted a 3 kb region that removed the three known TREs as well as the PRE. Surprisingly, this deletion had no effect. The authors suggested that this region does not contain TREs since no anterior transformations were observed. We rather suggest that, as trxG mutations partially complement PcG mutations, simultaneous deletion of these TREs along with their neighboring PRE complement the loss of the PRE.

PREs from the *Abdominal-B* (*Abd-B*) and the *Ubx cis*-regulatory sequences are nuclease hypersensitive[72,73] and nuclease-hypersensitive regions closely coincide with PRE activity in transgenic assays.[74] Furthermore, nuclease mapping and quantitative Chromatin Immunoprecipitation (ChIP) assays suggest that PREs are devoid of nucleosomes and depleted for histones.[75] A recent study using scanning force microscopy combined with DNA topological assays suggests that the *bxd* PRE is wrapped around a complex made up of Pho and the Polycomb core complex (PC, PH, PSC and dRing).[76] This negatively supercoiled structure would prevent the formation of nucleosomes. The authors propose that like promoters and enhancers, PREs would be in a nucleosome-depleted conformation in vivo.

A recent genome-wide analysis of PcG binding showed that at more than 200 PcG target genes, binding sites for the three PcG proteins PC, PSC and E(Z) colocalize to presumptive PREs.[67] However, PcG proteins also bind very large genomic regions from several to hundreds of kb[77,78] suggesting that they might spread out from PREs into flanking sequences, interact with histones and modify them. It is also possible that the ability of PRC1 to antagonize nucleosome remodelling can occur at a distance and that this activity requires the PRE. Conversely, the ETP GAF binds to narrow chromatin regions usually close to promoters.[77] Furthermore, the developmental binding profile of PH and PC is highly dynamic.[77,79]

Despite the extensive characterization of MEs and the proteins that bind them, the role of MEs remains unclear. In flies, MEs are required continuously throughout development. If inducible recombination is used to remove the *Miscadastral pigmentation* (*Mcp*) PRE in a transgene, derepression of the target locus in imaginal discs is evident with a few cell cycles.[80] Similar results are observed if PcG proteins are removed in mitotic clones.[81]

In Drosophila, PREs silence more effectively when they are paired, termed "pairing-sensitive repression".[69,82-84] The pairing assay shows that intra- and interchromosomal interactions between PREs are possible.[85] One way to explain this observation is that interactions among PcG proteins bring about interactions among PREs. Consistent with this idea, several groups noticed that transgenes containing PREs have a tendency to insert in the vicinity of their chromosomal origin, a phenomenon called "homing", possibly because of PcG protein interactions.[86] Interactions between PcG proteins bound to PREs may also account for the observations that antibodies to PcG proteins detect large, discrete bodies in nuclei termed "PcG Bodies".[82,87] Consistent with the idea that PREs interact at a distance, different PREs from the BX-C colocalize in PcG bodies in vivo.[70] It may be that PcG bodies sequester silenced target genes into a particular nuclear subcompartment where PcG or trxG proteins are abundant, or that pairing or association of PRE in PcG bodies creates aggregates of PcG proteins necessary to propagate the silent state.

Recent in vivo evidence shows that MEs can interact with distant DNA sequences.[70] Chromatin conformation capture was used to assay the three dimensional structure of PREs from the BX-C.[70] Strikingly the PREs and the promoters of the three coding genes colocalized in vivo. The authors propose that the PREs and coding promoters adopt a stable multi-looped structure that is organized in three domains that correspond to each coding region. Depletion of PcG proteins by RNA interference decreased the long-range interactions monitored by chromatin conformation capture, suggesting that these interactions are mediated by PcG proteins. Together, these experiments suggest that PREs have an important role in the generation of three dimensional structures in the nucleus.

Recruitment of Maintenance Proteins to Maintenance Elements

Most MPs do not bind to DNA but are thought to be recruited to chromatin via sequence specific factors as well as preferential binding to histone modifications. Apparently there are multiple ways to recruit MPs to MEs because there is little sequence conservation among MEs.

One model in the literature is that binding of the only specific DNA-binding proteins of the PcG, Pho (or PhoL) recruits PRC2, which methylates H3K27, which in turn recruits PRC1 for silencing.[34,88] This model is supported by some observations. Multiple consensus binding sites for Pho are indeed found within MEs.[47] The mammalian homolog of Pho is the transcription factor Yin Yang-1 (YY1) and multiple YY1 binding sites are also found in *Hoxa9* in mice.[89] In mammals there is evidence that YY1 recruits PRC2, as knockdown of YY1 with RNAi leads to loss of E(Z)H2 and H3K27 trimethylation.[90] Furthermore, Drosophila PC binds preferentially to H3K27me3[34] and PcG binding mostly coincides with H3K27me3 on genome-wide profiles.[78,91] Papp and Muller found K27 trimethylation as well as PRC1 binding throughout the regulatory and coding regions of *Ubx* in imaginal wing discs where *Ubx* is off.[45] Strong overlap of PC and H3K27me3 have also been observed on polytene chromosomes by Ringrose et al,[92] who observed no overlap between PC and H3K27me2.

Although this model is attractive, a number of more recent observations indicate that it is too simplistic. First, there are evidence that members of PRC1 also directly interact with Pho[62,93] and in Drosophila cells, PC binding to the promoter of *Ubx* occurs independently of E(Z) and Pho.[94] Secondly, Pho binding is unlikely to completely account for recruitment of PcG proteins to MEs since in Drosophila double-mutant *pho/PhoL*, binding of PcG proteins to larval polytene chromosomes is unaffected.[48] Deletion of the Pho consensus site from a transgene containing the *engrailed* PRE prevents pairing-sensitive repression, but so does deletion of other sequences. Deletion of Pho sites from an endogenous *bxd* PRE causes minor phenotypes in comparison with other sequences.[47,95,96] In an elegant series of experiments, Pirrotta's lab determined the effect of tethering different PcG proteins to a reporter in wild-type and PcG mutant backgrounds.[62,93] Tethered Pho is neither sufficient for silencing nor for interacting with PC-containing complexes. In contrast, tethering of PRC1 subunits (PC, PH, PSC) or the PRC2 subunit Su(Z)12 is sufficient to establish transient silencing, but this is not maintained later in development. If PRC2 is only required for PRC1 recruitment, it is surprising that mutations in *esc* prevent silencing by tethered PC. Thirdly, it is unlikely that H3K27 methylation is the sole determinant of PRC1 binding. In mouse ES cells, loss of EED, which results in complete loss of H3K27 methylation, results in upregulation of only 87% of PcG targets examined.[97] Human PC1 binds to trimethylated H3K27 in vitro, but not when trimethylation of K9 is also present and binds to H3K27me2 with greater affinity than it binds to H3K9me3.[98] But of the five murine Polycomb homologs, three bind to H3K9me3, either as well as to H3K27me3 (Cbx2 and Cbx7) or preferentially (Cbx4) and two do not bind either (Cbx6 and Cbx8).[99]

Other transcription factors have been implicated in recruitment of PcG proteins to ME. GAF and Pipsqueak (PSQ) recognize and bind d(GA)n repeats within MEs, facilitate silencing via their consensus sequences[74,100-102] and interact with PcG proteins and facilitate their binding to MEs. DSP1 recognizes G(A) sites within MEs and mediates PcG-dependent silencing.[65] A minimal PRE containing Pho consensus sites is unable to recruit Pho without the addition of the DSP1 consensus sequence, however deletion of an endogenous DSP1 site in a *bxd* PRE has minimal effects on silencing and does not enhance loss of silencing due to loss of Pho consensus sites. Nevertheless, with the exception of $(GA)_n$ sequences and Pho binding sequences, other binding sequences are not highly conserved among MEs, or necessary in vivo.[71]

Furthermore, binding of MPs to chromatin cannot be completely sequence-dependent because different MPs bind to different chromosomal regions depending on tissue type, developmental timing and transcriptional status of target genes.[77,103] However, two temporally and spatially regulated sequence specific transcription factors, Grainy head (GrH) and Sp1/KLF, are required for PcG recruitment to specific PREs, though whether they mediate temporal or spatial changes in PcG recruitment is unknown.[104,105]

Lastly, hierarchical recruitment models are insufficient to explain maintenance because some sequence specific factors can recruit both PcG and trxG complexes and thus might be considered ETPs. *Trl* was first identified as a trxG gene and was implicated in activation.[106] Pho, defined genetically as PcG, interacts with BRM as well as PC.[107] Zeste also interacts with trxG proteins,[108] but it is the only DNA binding protein that has been found as a stoichiometric component of PRC1[109] and it can mediate both activation and repression.[110,111] It is yet to be determined what regulates whether PcG or trxG proteins are recruited to a particular locus by sequence specific factors.

Role of Maintenance Proteins in Regulation of Transcription

PcG and trxG proteins also bind near transcription start sites, suggesting a role in regulation of transcription. The E(Z) protein trimethylates histone H3K27, which is thought to be a signal for silenced genes. But how does this bring about silencing? One clue is offered by observations that TBP and TAFs associate with PRC1 subunits,[109,112] consistent with the possibility that they interfere with transcriptional initiation. This idea has been tested directly using a transgene to observe the effect of PcG silencing on the *hsp26* promoter.[113] The presence of PcG proteins did not prevent recruitment of TBP, RNA polymerase II (Pol II) or the Heat Shock Factor to the transgene, arguing that chromatin structure does not prevent recruitment. However, promoter clearance, marked by the synthesis of a short strand of RNA and melting of the DNA ahead of Pol II, did not occur, suggesting that PcG proteins prevent it.

The Facilitates Chromatin Transcription (FACT) complex facilitates elongation by binding and displacing H2A/H2B dimers, therefore allowing Pol II to progress through chromatin (for review see 114). Zhou et al[115] have shown that the Spt16 component of FACT preferentially co-immunoprecipitates with non-ubiquitinated H2A over the ubiquitinated form and propose that ubiquitination of H2A (H2Aub) could prevent recruitment of the FACT complex to promoters. Consistent with a role for H2Aub in preventing elongation, Stock et al[116] showed that loss of both Ring1A and Ring1B resulted in loss of H2Au and derepression of target genes. These experiments do not exclude the possibility that roles for Ring1A and Ring1B outside of their ubiquitin E3 ligase enzymatic activity account for the results (discussed in 117).

Ubiquitination of H2A leads to binding by Ring and YY1 binding Protein (RYBP)[118] and transcriptional repression by RYBP is dependent on the presence of PRC1.[119] Furthermore, the E2F transcription factor family member E2F-6 may be involved in stable repression of the Hox genes via RYBP. The marked box domain of E2F6 interacts with the N-terminal of RYBP.[120] The most striking phenotype of *E2f6* knockout mice are posterior transformations of the 6th lumbar to the 1st sacral vertebra and of the 13th thoracic to the 1st lumbar vertebra,[121] suggesting that it contributes to Hox gene regulation. E2F6 is a member of multiple complexes containing PcG members. First, it interacts with the PcG proteins Ring1, Bmi-1, MEL-18 and mph1.[120] Ogawa et al[122] identified Ring1 and Ring2 as part of another E2F6 complex involved in silencing of E2F- and Myc-responsive genes in quiescent cells. Lastly, E2F6 also belongs to a complex containing DP1, EPC1 and Sin3b that interacts with EZH2.[123] This complex have been implicated in the regulation of genes involved in cell cycle progression.[123] Mutations in *E2f6* enhance the skeletal homeotic defects of *Bmi-1* knockout mice, but not the cell cycle defects attributed to misregulation of *Ink4a*, raising the possibility that the Bmi1-E2F6 complex regulates Hox gene but not cell cycle gene expression.[121]

The trxG proteins are important for more than one phase of transcription. In yeast the Trx homologs Set1 and Set2 methylate H3K4 and H3K36, respectively. H3K4 methylation is highest near promoters and is required for transcriptional initiation and promoter clearance, as well as early transcriptional elongation. Consistent with the pattern of H3K4 methylation, Trx binding is highest near promoters,[124,125] suggesting that Trx also plays a role in transcriptional activation. However, Trx is also required for transcriptional elongation. Heat shock genes have paused Pol II that has completed promoter clearance but is blocked from attaining transcriptional elongation. Mutations in *trx* prevent the heat shock response and prevent recruitment of factors required for elongation.[125] Trx has also a role in transcriptional elongation at the *Ubx* locus because mutations do not prevent transcription of RNA at the promoter, but severely decrease full-length transcripts.[124] H3K36 methylation is required for multiple rounds of transcriptional elongation[126] and is found at equivalent levels throughout the transcription unit suggesting that ASH1 has also a role in elongation. Correspondingly, ASH1 binding in mammals is found at equivalent levels throughout the transcription unit.[127] Contrary to these molecular observations, genetic evidence has been used to argue that trxG proteins do not have a role in transcriptional initiation or elongation.[128] In imaginal discs, Hox expression is lost in *ash1* or *trx* mutants, but is restored in *ash1* or *trx* mutants that also lack PcG function. Furthermore, Hox genes are ectopically expressed in discs mutant for PcG genes, even if they are also mutant for trxG genes. These data are consistent with the idea that trxG proteins are anti-repressors that antagonize the silencing effect of PcG proteins rather than activators of transcription. Controversy between these molecular and genetic results could be due to redundancy between ASH1, Trx and other trxG proteins such as Kismet, involved in an early step in transcriptional elongation.[129] Alternatively, trxG could relieve PcG-caused RNA Pol II block.

How ETPs promote either the activation or the repression of a defined target gene remains an open question. One explanation could be that the phenotypes arise from a cascade of interactions. Moreover, PcGs and ETPs positively regulate each other's expression[130] and there are also complex cross-regulations between Hox genes. On the other hand, some ETPs like *Asx* in Drosophila exhibit tissue-specific effects.[13] This feature is conserved in mammals since *Asxl1*^{tm1Bc} mutants show derepression of *Hoxc8* in posterior ectoderm and mesoderm whereas they show reduced activation of *Hoxc8* in the brain. In human HeLa cells, mouse Asxl1 tethered to GAL4 binding sites activates a heterologous thymidine kinase promoter driving a luciferase reporter and also increases RAR-dependent-activation of the reporter in transient transfection assays. However, in other mammalian cell types, Asxl1 acts as a RAR corepressor.[131] The ETP *corto* also exhibits a pair-rule expression pattern in early embryos.[132] Thus, we propose that ETPs exhibit context-dependent activity, depending on the proteins with which they interact and the regulatory sequences which are present. Their function would be determined by their interacting partners or chromatin environment to bring about changes in gene regulation.

Several studies have recently called into question the simple model that PcG binding versus trxG binding determines transcriptional silencing versus activation, respectively. Genome-wide ChIP analysis show that 10-20% of PcG target genes are actively transcribed in human and mouse ES cells, Drosophila *Sg4* and *Kc1* cells and human embryonic fibroblasts (reviewed in ref. 67). In Drosophila *S2* cells, Ringrose et al[133] observed no correlation with either Pc binding, H3K27me3, or H3k9me3 and transcription as assayed by RT-PCR. Furthermore, binding of trxG and PcG proteins to the *Ubx* gene in its off versus on state has been examined in Drosophila larvae.[45] PhoRC, PRC1, PRC2 and Trx are present on the *bxd* and *bx* PREs whatever the gene is transcribed or not suggesting that recruitment of PcG complexes to PREs occurs by default. In the off state, the trimethylation marks H3K27me3, H3K9me3 and H4K20me3 extend over a 100-kb domain that spans the whole *Ubx* gene whereas in the on state these silencing marks are restricted to regions upstream of the promoter being replaced downstream by H3K4me3. ASH1 binds to the promoter and downstream of the transcription start only in the on state. It could prevent methylation of the promoter and coding region by PRC2 when *Ubx* is activated. Moreover, in the on state, the H3K27me3 mark is reduced at the PRE suggesting that PREs could serve as assembly platforms for PcG complexes like PRC2 that act through large distances to stamp chromatin on vast domains. Conversely, Petruk et al[66] show in an elegant study that in salivary glands of third instar Drosophila larvae, at the single-cell level, binding of PcG and trxG proteins to the *bxd* PRE are mutually exclusive and correlates with the activated or repressed state of an *Ubx* transgene, respectively. Single cell studies have an advantage over multi-cell ChIP studies in that transcriptional states and therefore binding may not be uniform throughout a cell population. However polytene tissues may be different than diploid tissues.

Epigenetic Marks

The effects Maintenance Proteins exert on transcription are stable to cell division and therefore epigenetic. Epigenetic marks that are stable to DNA synthesis and mitosis have been proposed as a mechanism for the maintenance of epigenetic memory but their identity is not known.

DNA methylation provides a clear example of a molecular mark that is stable to DNA synthesis and mitosis and that can be propagated after semi-conservative replication owing to the existence of DNA hemi-methylase DNMT1.[134] Furthermore, in mammals, DNA methylation could play a role in PcG-mediated silencing. EZH2 interacts with DNMT1, DNMT3a and DNMT3b, EZH2 knockdown causes loss of DNMT binding and loss of DNA methylation at PcG targets.[135]

Disruption of DNMT1 causes improper recruitment of Bmi1 and Ring1B to PcG bodies.[136] During in vitro ES cell differentiation, 21% of genes that loose H3K4me2 and maintain H3K27me3 are de novo DNA methylated and PRC2 targets are more likely than nonPRC2 targets to be de novo methylated.[137] A complex containing the maintenance methyltransferase DNMT1 and PRC1 members Bmi1 and Ring1 has been observed.[138] The transcriptional repressor Dmap1 colocalizes with DNMT1 and HDAC2 at replication foci and localizes to pericentric heterochromatin during late S phase, presumably to promote DNA methylation as well as histone deacetylation and transcriptional silencing of newly replicated DNA.[139] Binding of Dmap1 and DNMT1 to certain genes is lost in *Bmi1* mutant cells,[138] but another report has shown that localization of Bmi1 to PcG bodies depends on DNMT1.[136] An attractive hypothesis is that PRC1 accompanies the maintenance methyltransferase at the replication fork to promote transcriptional silencing of newly synthesized daughter strands. Thus, DNA methylation may serve as a memory mark in vertebrates and simply stabilize silencing which is brought about by other mechanisms.[134] Maintenance proteins are well conserved between flies and humans, but DNA methylation patterns are not. In Drosophila, cytosine methylation occurs very rarely in the context of CpG dinucleotides and therefore is nonsymmetrical.[140] In this context, no mechanism for inheritance of methylation through DNA synthesis is known. There is however some evidence for CpG methylation playing a role in transcriptional silencing and possibly in PcG-dependent silencing in Drosophila. A screen for genes involved in tumorigenesis identified a double mutation in *psq* and *lola* that is associated with downregulation of *Rbf* accompanied by increased promoter CpG methylation. Importantly, mutations in *E(z)*, *Esc* and *Pc* suppressed tumor formation in *psq lola* double mutants.[141]

Transcription factors themselves may be epigenetic, possibly without the involvement or requirement for another mark, due to cytoplasmic inheritance through mitosis.[3] However, many transcription factors are expressed only transiently, as is the case for regulators of Hox expression. It is not obvious how they could be replenished to account for the dilution caused by cell division. The same objection holds for transcription factors that remain bound to DNA. However, some DNA-binding proteins required for PcG or trxG recruitment, such as GAF or PSQ, remain on chromosomes during mitosis.[142] Ubiquitous transcription factors could function as marks in conjunction with maintenance proteins or histone modifications. For example, RYBP binds to ubiquitinated proteins in vivo and in vitro and colocalizes with H2Aub and Ring1b on chromatin.[118] Since transcriptional repression by RYBP depends on PRC1,[119] binding of RYBP to H2Aub following S phase could recruit PRC1 and allow for the propagation of the modification onto the newly synthesized histones.

Histone modifications have also been proposed to be epigenetic. They can be stable to multiple rounds of cell division and sometimes to nuclear transplantation and reprogramming. It is possible that conservative histone inheritance could allow for propagation of histone modifications if histone modifying enzymes are preferentially recruited to chromosomal regions containing the mark they are responsible for. This could explain phenomena such as the "spreading" of heterochromatin observed at Drosophila centromeres as well as the tendency for histone modifications to exist in large chromosomal domains. However, in some cases, histone modifications are not stably maintained through cell division and therefore are not epigenetic. H3K27me3 decreases as cells progress through S phase in a human cell line and in yeast H3K9me at some genomic regions is not propagated through DNA synthesis.[136,143] Moreover, loss of E(Z) in flies leads to gradual loss of H3K27 methylation, implying that H3K27 methylation is not sufficient to be the epigenetic mark.[88]

Histone variants could be the epigenetic mark. Most research has focused on the histone H3 variants. Most organisms have a centromere specific H3 variant CID (CENP-A/Cse-4) that is essential for mitosis, as well as one or more H3 variants (reviewed in ref. 144). The variant H3.3 is enriched with methylated K4 and K79 and acetylated K9, K14, K18 and K23, modifications which are associated with transcriptional activation. On the contrary, H3.2 is enriched with modifications associated with silencing, such as methylations of K9 and dimethylation or trimethylation of K27.[145] Mammalian cells also contain the variant H3.1, which is enriched with both active and silent marks, K14 acetylation and K9 dimethylation, respectively.[145] *Saccharomyces cerevisiae*, which lacks heterochromatin marks such as K9 and K27 methylation, has only one variant, H3.3 and *Schizosaccharomyces pombe* contains one variant with similarities to both H3.3 and H3.2. Another reason for interest in the histone variants as epigenetic arises from the observation that H3.3 deposition is replication independent while H3.1 deposition is replication dependent.[146] This difference is mediated by two chaperones, the replication-dependent chaperone CAF-1, which is localized to the replication fork via an interaction with PCNA,[147] as well as the replication-independent chaperone HIRA.[148] Various models have been proposed relating histone variants to the epigenetic inheritance of transcriptional states.[144,149] Recent nuclear transplant experiments have illustrated the stability and importance of histone variants in the maintenance of cellular differentiation. Ng and Gurdon[150] examined the transcription state and chromatin structure of the *MyoD* gene following nuclear transplantation of somite nuclei, which express *MyoD*, into an enucleated egg. The transcription state of *MyoD* can be maintained for 24 cell divisions, independent of the DNA methylation state of the promoter. Instead, memory is correlated with the presence of H3.3 at the promoter. Memory cannot be maintained when H3.3K4 cannot be methylated. However, these experiments fail to demonstrate that H3.3 variants are actively propagated and are not diluted through cell division.

It is possible that MPs themselves could be the epigenetic mark. The DNA replication processivity factor PCNA that participates in the inheritance of chromatin structure during S phase in yeast,[147] colocalizes with the PcG protein Cramped in Drosophila,[151] suggesting that PcG proteins may have a direct role in promoting inheritance of their targets possibly by localizing to the replication fork. Binding of some PRC1 members is not stable to mitosis,[87] suggesting that they are not the epigenetic mark. Nevertheless, a recent report suggests that some PcG and trxG proteins are stable constituents of mitotic chromosomes.[152] Perhaps the small amount of MP proteins that do remain bound to chromatin through mitosis could be sufficient to serve as a mark even if it is not readily detected by immunohistochemistry. Consistent with this possibility, binding of PRC2 members EED and EZH2, as well as PRC1 member Ring1b, to the inactive X chromosome in TS cells is not lost during mitosis.[153,154]

However, there are several reasons to argue that MPs are not the epigenetic mark. As explained above, several studies show that a model in which PcG binding versus trxG binding determines transcriptional silencing versus activation is too simplistic. Furthermore, the emerging evidence of the reversibility of PcG mediated silencing also calls into question simple models that MPs and their modifications are stable epigenetic marks. Many observations indeed show that PcG silencing is reversible. First, a H3K4me3 demethylase has been discovered which is recruited by PRC2 during ES cell differentiation and was originally identified as a trxG gene in Drosophila.[155] It also appears that the binding of PRC1 to chromatin is not stable but is in equilibrium with the soluble fraction[156] and PcG mediated silencing of lineage-specific genes in embryonic stem cells is reversed during differentiation of ES cells.[97,157]

Release of PcG Silencing

Experiments in which cells of imaginal discs change fate ("transdetermination") show that change in gene expression patterns requires a change that occurs in S phase.[158] This link between S phase and change in maintenance has been observed in other systems including neuronal cells[159] and vulva precursors in *Caenorrhabditis elegans*.[160] Disruption of the cell cycle upsets Hox regulation in chick somites.[161] Intriguingly, the frequency of transdetermination is altered in PcG and trxG mutants,[162,163] perhaps because of their role in chromatin remodeling.[164]

Replication patterns in the mammalian PcG target locus *HoxB* have been studied in relation to transcriptional status of the region. The *HoxB* domain is not transcribed in undifferentiated mouse P19 cells and replication origin use is unspecified and spread throughout the locus.[165] Following retinoic acid (RA)-induced differentiation of the cells and activation of the locus, which mimics temporal colinearity of activation during murine development, replication is restricted to a single origin upstream of the *HoxB1* gene and at the edge of the *HoxB* domain.[165]

Interestingly, the pattern of histone modifications that accompany specification of this origin are counter-intuitive. As expected the levels of acetylated histones H3 and H4 increase following RA treatment, however not in the location of the origin.[165] Therefore, origin use is decreased in regions where histone acetylation is increased and vice-versa, which is in opposition to the view that replication origins are correlated with histone acetylation and transcriptional activity (reviewed in ref. 166). Further, the use of the origin upstream of the *HoxB1* gene is not linked to expression of this particular gene, as it remains in use 3 days after RA induction at which time the gene is no longer active.[165] It would be interesting to determine if and how developmental changes in origin use and timing contribute to maintenance or are dependent on MPs. In Xenopus embryos, *Hoxb3-Hoxb9* expression does not occur if DNA replication is blocked before the mid-blastula transition. It is likely that when fate change occurs, nascent DNA allows for the formation of a new chromatin template that is permissive to transcriptional activity.[167]

Role of PcG Proteins in Chromatin Replication

As DNA replication is the only time during which the whole genome is assayed and the epigenetic mark must be stable to DNA synthesis, many researchers have tried to establish a link between PcG genes and DNA synthesis.

In Drosophila, the gene amplification that occurs in follicle cell nuclei has been studied extensively as a model for replication origin specification and initiation. During replication initiation, histone H4K5, K8 and K12 as well as H3K14 are acetylated and this is lost in *dup/cdt1* mutants.[168] Loss of the histone deacetylase Rpd3 causes increase in global acetylation levels as well as increased origin usage outside of normally amplified regions which was not blocked by treatment with the transcriptional inhibitor α-aminitin, indicating the effect is not indirect.[6] Further, tethering of PC to the amplified origin inhibited origin activity, presumably due to effects on chromatin structure.[6]

Rae28, a mammalian PH homolog, interacts with the replication licensing factor Geminin.[169] Geminin inhibits Cdt1/Dup thereby inhibiting formation of the Pre-Replication Complex.[170] The interaction between Rae28 and Geminin likely mediates poly-ubiquitination of Geminin therefore its degradation.[171] Interestingly, over-expression of *geminin* inhibits *Hoxb9* transcription in chick embryos.[169] However, the effects of Geminin on *Hoxb9* expression could be independent from its role in DNA synthesis since during neurogenesis Geminin promotes maintenance of the undifferentiated state by antagonizing the SWI/SNF chromatin remodeling protein Brg1.[172]

Role of PcG Proteins in Stem Cells

Bmi-1 is critical for haematopoietic stem cell (HSC) self-renewal. Expression levels are high in mouse and human HSCs and *Bmi-1* mutant have reduced levels of HSCs.[173] Transplantation of *Bmi-1*$^{-/-}$ fetal liver cells into irradiated mice causes only a transient reconstitution of myeloid, B- and T-cells, indicating loss of HSC self-renewal.[173,174] As well as the *ink4a* locus, *Bmi-1*$^{-/-}$ HSCs have upregulated levels of the p53-induced gene *wig1* and downregulation of the apoptosis inhibitor *AI-6*. This indicates that Bmi-1 induces apoptosis as well as senescence, downregulating an apoptotic inhibitor as well as activating the p53 pathway through the *ARF* locus. Bmi-1 may also contribute to extension of the replicative lifespan and immortalization through upregulation of the human telomerase RT gene (hTERT).[175]

Bmi-1 is also critical to neural stem cell self-renewal, which are postnataly depleted in *Bmi-1* deficient mice, correlating with an increased expression of *p16*^{ink4a} RNA.[176] Interestingly, some neural progenitors proliferate normally, distinguishing proliferation from self-renewal. HSCs transfected with *Hoxa9* and *Meis1a* are capable of producing acute myelogenous leukemia (AML), but transfected cells derived from *Bmi-1*$^{-/-}$ mice show lower levels of leukemic cells in the blood and HSCs from primary recipients are not capable of causing AML in secondary recipients.[174]

EED, Suz12 and *Rae28/HPH1,* are also necessary for stem cell maintenance (reviewed in ref. 177). The emergence of multiple genome wide ChIP experiments in differentiated as well as ES cells in both humans and mice allowed for the creation of a new model for the role of PcG proteins in stem cell maintenance. One key finding from genome wide ChIP studies was the identification of key differentiation regulators as PcG targets in both human and mice ES cells.[97,157] Repression of transcription factors known to induce differentiation such as *Sox, Gata* and *Tbx* by the PcG supports a role for them in maintenance of stem cells in the undifferentiated state. However binding of PcG proteins to these regions implies a requirement for PcG-dependent silencing to be reversible.

Analysis of chromatin modifications found on genes repressed by PcG proteins revealed the presence of "bivalent domains" which contain the PRC2 mediated H3K27me3, but also the activating modification H3K4me3.[178,179] During differentiation these domains resolve to contain only the activating or repressive mark, no mark and a small percentage remain bivalent.[180] There is some evidence that the levels of bivalency in a cell type is correlated with its differentiation potential[180,181] and it will be interesting to determine whether this explains the requirement for PcG proteins in maintenance of adult stem cells as well.

Conclusion

Enormous progress has been made in recent years in understanding molecular aspects of PcG and trxG function. The advent of ChIP, genome-wide ChIP, has shown where and in some cases when PcG and trxG proteins bind in genomes. Purification of multisubunit complexes and analysis of subunit interactions and function of the overall complex has paved the way to understanding the roles of these proteins in chromatin structure and transcription.

Future Research in the Field

Some major questions that remain to be answered. There is increasing appreciation that the role of PcG and trxG proteins is dynamic and changes throughout development. As a consequence, static models of PcG or trxG function must be modified to explain how these groups of proteins can bind targets but not function in some cases, or function in other cases. The discoveries that histone methylation is reversible and that multivalent signals have an important role in a histone code suggest that it will be important to look beyond H3K4 and H3K27 methylation and to assess the role of cofactors, perhaps ETPs in mediating PcG and trxG function. Most studies look at function after maintenance has been achieved. Further advances will require a system in which dynamic changes in PcG and trxG function can be monitored and experimentally altered. In such a system, high resolution studies of binding, histone modification and analysis of cofactors in wild-type and mutant backgrounds will be required to unravel how these proteins affect transcription.

Up to now, most attention has been paid to identifying ME and understanding how ME recruit PcG and trxG proteins. While there is some understanding of the role of PcG and trxG proteins at promoters, their role at ME is very poorly defined. Furthermore, it is not clear whether there are discrete ME in mammals. In Drosophila, the role of ME in maintenance is not understood. In the future, we expect to see experimental analysis of the roles of ME-promoter interactions, pairing of ME, roles of ME in formation of PcG bodies or the establishment of nuclear architecture. It may be that the role of ME is related to DNA replication timing and that its role in silencing is moderated this way.

Curiously, little attention has been paid to the key feature of trxG and PcG proteins, namely that they are epigenetic. We know essentially nothing about the identity and propagation of epigenetic marks in maintenance. The roadblock has been the difficulty of characterizing newly replicated DNA and the proteins bound to it in vivo. Much has been learned from reconstituted DNA replication systems, but there is a dearth of in vivo studies. There are some considerable technical issues, as a specific short sequence will be replicated in a very small proportion of a population, so material will be limiting. Finding a way to answer this question is a challenge for the future.

References

1. Jacobs JJ, van Lohuizen M. Polycomb repression: from cellular memory to cellular proliferation and cancer. Biochim Biophys Acta 2002; 1602:151-161.
2. Cavalli G. Chromatin as a eukaryotic template of genetic information. Curr Opin Cell Biol 2002; 14:269-278.
3. Ptashne M. On the use of the word 'epigenetic'. Curr Biol 2007; 17:R233-236.
4. Sparmann A, van Lohuizen M. Polycomb silencers control cell fate, development and cancer. Nat Rev Cancer 2006; 6:846-856.
5. Brock HW, Fisher CL. Maintenance of gene expression patterns. Dev Dyn 2005; 232:633-655.
6. Aggarwal BD, Calvi BR. Chromatin regulates origin activity in Drosophila follicle cells. Nature 2004; 430:372-376.
7. Gildea JJ, Lopez R, Shearn A. A screen for new trithorax group genes identified little imaginal discs, the Drosophila melanogaster homologue of human retinoblastoma binding protein 2. Genetics 2000; 156:645-663.
8. Lewis EB. A gene complex controlling segmentation in Drosophila. Nature 1978; 276:565-570.
9. Struhl G, Akam M. Altered distributions of Ultrabithorax transcripts in extra sex combs mutant embryos of Drosophila. EMBO J 1985; 4:3259-3264.
10. Duncan IM. Polycomblike: a gene that appears to be required for the normal expression of the bithorax and antennapedia gene complexes of Drosophila melanogaster. Genetics 1982; 102:49-70.
11. Slifer EH. A mutant stock of Drosophila with extra sex combs. J Exp Zool 1942; 90:31-40.
12. Pattatucci AM, Kaufman TC. The homeotic gene Sex combs reduced of Drosophila melanogaster is differentially regulated in the embryonic and imaginal stages of development. Genetics 1991; 129:443-461.
13. Soto MC, Chou TB, Bender W. Comparison of germline mosaics of genes in the Polycomb group of Drosophila melanogaster. Genetics 1995; 140:231-243.
14. Nusslein-Volhard C, Kluding H, Jurgens G. Genes affecting the segmental subdivision of the Drosophila embryo. Cold Spring Harb Symp Quant Biol 1985; 50:145-154.
15. Landecker HL, Sinclair DA, Brock HW. Screen for enhancers of Polycomb and Polycomblike in Drosophila melanogaster. Dev Genet 1994; 15:425-434.
16. Sato T, Russell MA, Denell RE. Homoeosis in Drosophila: A New Enhancer of Polycomb and Related Homeotic Mutations. Genetics 1983; 105:357-370.
17. Wu CT, Jones RS, Lasko PF et al. Homeosis and the interaction of zeste and white in Drosophila. Mol Gen Genet 1989; 218:559-564.
18. Gaytan de Ayala Alonso A, Gutierrez L, Fritsch C et al. A genetic screen identifies novel polycomb group genes in Drosophila. Genetics 2007; 176:2099-2108.
19. Ingham PW, Whittle R. Trithorax: a new homeotic mutation of Drosophila melanogaster causing transformations of abdominal and thoracic imaginal segments. I. Putative role during embryogenesis. Mol Gen Genet 1980; 179:607-614.
20. Mazo AM, Huang DH, Mozer BA et al. The trithorax gene, a trans-acting regulator of the bithorax complex in Drosophila, encodes a protein with zinc-binding domains. Proc Natl Acad Sci USA 1990; 87:2112-2116.
21. Ingham PW. Differential expression of bithorax complex genes in the absence of the extra sex combs and trithorax genes. Nature 1983; 306:591-593.
22. Kennison JA. Transcriptional activation of Drosophila homeotic genes from distant regulatory elements. Trends Genet 1993; 9:75-79.
23. Kennison JA. The Polycomb and trithorax group proteins of Drosophila: trans-regulators of homeotic gene function. Annu Rev Genet 1995; 29:289-303.
24. Milne TA, Sinclair DA, Brock HW. The Additional sex combs gene of Drosophila is required for activation and repression of homeotic loci and interacts specifically with Polycomb and super sex combs. Mol Gen Genet 1999; 261:753-761.
25. Bejarano F, Busturia A. Function of the Trithorax-like gene during Drosophila development. Dev Biol 2004; 268:327-341.
26. Lopez A, Higuet D, Rosset R et al. corto genetically interacts with Pc-G and trx-G genes and maintains the anterior boundary of Ultrabithorax expression in Drosophila larvae. Mol Genet Genomics 2001; 266:572-583.
27. Kodjabachian L, Delaage M, Maurel C et al. Mutations in ccf, a novel Drosophila gene encoding a chromosomal factor, affect progression through mitosis and interact with Pc-G mutations. EMBO J. 16 1998; 17:1063-1075.
28. Decoville M, Giacomello E, Leng M et al. DSP1, an HMG-like protein, is involved in the regulation of homeotic genes. Genetics 2001; 157:237-244.
29. Schuettengruber B, Chourrout D, Vervoort M et al. Genome regulation by polycomb and trithorax proteins. Cell 2007; 128:735-745.

30. Schwartz YB, Pirrotta V. Polycomb silencing mechanisms and the management of genomic programmes. Nat Rev Genet 2007; 8:9-22.
31. Wang H, Wang L, Erdjument-Bromage H et al. Role of histone H2A ubiquitination in Polycomb silencing. Nature 2004; 431:873-878.
32. Francis NJ, Saurin AJ, Shao Z et al. Reconstitution of a functional core polycomb repressive complex. Mol Cell 2001; 8:545-556.
33. Francis NJ, Kingston RE, Woodcock CL. Chromatin compaction by a polycomb group protein complex. Science 2004; 306:1574-1577.
34. Fischle W, Wang Y, Jacobs SA et al. Molecular basis for the discrimination of repressive methyl-lysine marks in histone H3 by Polycomb and HP1 chromodomains. Genes Dev 2003; 17:1870-1881.
35. Ketel CS, Andersen EF, Vargas ML et al. Subunit contributions to histone methyltransferase activities of fly and worm polycomb group complexes. Mol Cell Biol 2005; 25:6857-6868.
36. Muller J, Hart CM, Francis NJ et al. Histone methyltransferase activity of a Drosophila Polycomb group repressor complex. Cell 2002; 111:197-208.
37. Cao R, Wang L, Wang H et al. Role of histone H3 lysine 27 methylation in Polycomb-group silencing. Science 2002; 298:1039-1043.
38. Czermin B, Melfi R, McCabe D et al. Drosophila enhancer of Zeste/ESC complexes have a histone H3 methyltransferase activity that marks chromosomal Polycomb sites. Cell 2002; 111:185-196.
39. Ebert A, Schotta G, Lein S et al. Su(var) genes regulate the balance between euchromatin and heterochromatin in Drosophila. Genes Dev 2004; 18:2973-2983.
40. Martin C, Cao R, Zhang Y. Substrate preferences of the EZH2 histone methyltransferase complex. J Biol Chem 2006; 281:8365-8370.
41. Kuzmichev A, Margueron R, Vaquero A et al. Composition and histone substrates of polycomb repressive group complexes change during cellular differentiation. Proc Natl Acad Sci USA 2005; 102:1859-1864.
42. Nekrasov M, Wild B, Muller J. Nucleosome binding and histone methyltransferase activity of Drosophila PRC2. EMBO Rep 2005; 6:348-353.
43. Nekrasov M, Klymenko T, Fraterman S et al. Pcl-PRC2 is needed to generate high levels of H3-K27 trimethylation at Polycomb target genes. Embo J 2007; 26:4078-4088.
44. Ebert A, Lein S, Schotta G et al. Histone modification and the control of heterochromatic gene silencing in Drosophila. Chromosome Res 2006; 14:377-392.
45. Papp B, Muller J. Histone trimethylation and the maintenance of transcriptional ON and OFF states by trxG and PcG proteins. Genes Dev 2006; 20:2041-2054.
46. Klymenko T, Papp B, Fischle W et al. A Polycomb group protein complex with sequence-specific DNA-binding and selective methyl-lysine-binding activities. Genes Dev 2006; 20:1110-1122.
47. Brown JL, Mucci D, Whiteley M et al. The Drosophila Polycomb group gene pleiohomeotic encodes a DNA binding protein with homology to the transcription factor YY1. Mol Cell 1998; 1:1057-1064.
48. Brown JL, Fritsch C, Mueller J et al. The Drosophila pho-like gene encodes a YY1-related DNA binding protein that is redundant with pleiohomeotic in homeotic gene silencing. Development 2003; 130:285-294.
49. Savla U, Benes J, Zhang J et al. Recruitment of Drosophila Polycomb-group proteins by Polycomblike, a component of a novel protein complex in larvae. Development 2008; 135813-817.
50. Petruk S, Sedkov Y, Smith S et al. Trithorax and dCBP acting in a complex to maintain expression of a homeotic gene. Science 2001; 294:1331-1334.
51. Shilatifard A. Molecular implementation and physiological roles for histone H3 lysine 4 (H3K4) methylation. Curr Opin Cell Biol 2008; 20:341-348.
52. Papoulas O, Beek SJ, Moseley SL et al. The Drosophila trithorax group proteins BRM, ASH1 and ASH2 are subunits of distinct protein complexes. Development 1998; 125:3955-3966.
53. Beisel C, Imhof A, Greene J et al. Histone methylation by the Drosophila epigenetic transcriptional regulator Ash1. Nature 2002; 419:857-862.
54. Byrd KN, Shearn A. ASH1, a Drosophila trithorax group protein, is required for methylation of lysine 4 residues on histone H3. Proc Natl Acad Sci USA 2003; 100:11535-11540.
55. Tanaka Y, Katagiri Z, Kawahashi K et al. Trithorax-group protein ASH1 methylates histone H3 lysine 36. Gene 2007; 397:161-168.
56. Kuzin B, Tillib S, Sedkov Y et al. The Drosophila trithorax gene encodes a chromosomal protein and directly regulates the region-specific homeotic gene fork head. Genes Dev 1994; 8:2478-2490.
57. Rozovskaia T, Tillib S, Smith S et al. Trithorax and ASH1 interact directly and associate with the trithorax group-responsive bxd region of the Ultrabithorax promoter. Mol Cell Biol 1999; 19:6441-6447.
58. Rozenblatt-Rosen O, Rozovskaia T, Burakov D et al. The C-terminal SET domains of ALL-1 and TRITHORAX interact with the INI1 and SNR1 proteins, components of the SWI/SNF complex. Proc Natl Acad Sci USA 1998; 95:4152-4157.

59. Mohrmann L, Verrijzer CP. Composition and functional specificity of SWI2/SNF2 class chromatin remodeling complexes. Biochim Biophys Acta 2005; 1681:59-73.
60. Janody F, Martirosyan Z, Benlali A et al. Two subunits of the Drosophila mediator complex act together to control cell affinity. Development 2003; 130:3691-3701.
61. Salvaing J, Lopez A, Boivin A et al. The Drosophila Corto protein interacts with Polycomb-group proteins and the GAGA factor. Nucleic Acids Res 2003; 31:2873-2882.
62. Poux S, Melfi R, Pirrotta V. Establishment of Polycomb silencing requires a transient interaction between PC and ESC. Genes Dev 2001; 15:2509-2514.
63. Salvaing J, Decoville M, Mouchel-Vielh E et al. Corto and DSP1 interact and bind to a maintenance element of the Scr Hox gene: understanding the role of Enhancers of trithorax and Polycomb. BMC Biol 2006; 4:9.
64. Grimaud C, Negre N, Cavalli G. From genetics to epigenetics: the tale of Polycomb group and trithorax group genes. Chromosome Res 2006; 14:363-375.
65. Dejardin J, Rappailles A, Cuvier O et al. Recruitment of Drosophila Polycomb group proteins to chromatin by DSP1. Nature 2005; 434:533-538.
66. Petruk S, Smith ST, Sedkov Y et al. Association of trxG and PcG proteins with the bxd maintenance element depends on transcriptional activity. Development 2008; 135:2383-2390.
67. Ringrose L, Paro R. Polycomb/Trithorax response elements and epigenetic memory of cell identity. Development 2007; 134:223-232.
68. Simon J, Chiang A, Bender W et al. Elements of the Drosophila bithorax complex that mediate repression by Polycomb group products. Dev Biol 1993; 158:131-144.
69. Kassis JA. Pairing-sensitive silencing, polycomb group response elements and transposon homing in Drosophila. Adv Genet 2002; 46:421-438.
70. Lanzuolo C, Roure V, Dekker J et al. Polycomb response elements mediate the formation of chromosome higher-order structures in the bithorax complex. Nat Cell Biol 2007; 9:1167-1174.
71. Sipos L, Kozma G, Molnar E et al. In situ dissection of a Polycomb response element in Drosophila melanogaster. Proc Natl Acad Sci USA 2007; 104:12416-12421.
72. Dellino GI, Tatout C, Pirrotta V. Extensive conservation of sequences and chromatin structures in the bxd polycomb response element among Drosophilid species. Int J Dev Biol 2002; 46:133-141.
73. Barges S, Mihaly J, Galloni M et al. The Fab-8 boundary defines the distal limit of the bithorax complex iab-7 domain and insulates iab-7 from initiation elements and a PRE in the adjacent iab-8 domain. Development 2000; 127:779-790.
74. Mishra RK, Mihaly J, Barges S et al. The iab-7 polycomb response element maps to a nucleosome-free region of chromatin and requires both GAGA and pleiohomeotic for silencing activity. Mol Cell Biol 2001; 21:1311-1318.
75. Schwartz YB, Kahn TG, Pirrotta V. Characteristic low density and shear sensitivity of cross-linked chromatin containing polycomb complexes. Mol Cell Biol 2005; 25:432-439.
76. Mohd-Sarip A, van der Knaap JA, Wyman C et al. Architecture of a polycomb nucleoprotein complex. Mol Cell 2006; 24:91-100.
77. Negre N, Hennetin J, Sun LV et al. Chromosomal distribution of PcG proteins during Drosophila development. PLoS Biol 2006; 4:e170.
78. Schwartz YB, Kahn TG, Nix DA et al. Genome-wide analysis of Polycomb targets in Drosophila melanogaster. Nat Genet 2006; 38:700-705.
79. Kwong C, Adryan B, Bell I et al. Stability and dynamics of polycomb target sites in Drosophila development. PLoS Genet 2008; 4:e1000178.
80. Busturia A, Wightman CD, Sakonju S. A silencer is required for maintenance of transcriptional repression throughout Drosophila development. Development 1997; 124:4343-4350.
81. Beuchle D, Struhl G, Muller J. Polycomb group proteins and heritable silencing of Drosophila Hox genes. Development 2001; 128:993-1004.
82. Bantignies F, Grimaud C, Lavrov S et al. Inheritance of Polycomb-dependent chromosomal interactions in Drosophila. Genes Dev 2003; 17:2406-2420.
83. Vazquez J, Muller M, Pirrotta V et al. The Mcp Element Mediates Stable Long-Range Chromosome-Chromosome Interactions in Drosophila. Mol Biol Cell 2006; 17:2158-2165.
84. Muller M, Hagstrom K, Gyurkovics H et al. The mcp element from the Drosophila melanogaster bithorax complex mediates long-distance regulatory interactions. Genetics 1999; 153:1333-1356.
85. Sigrist CJ, Pirrotta V. Chromatin insulator elements block the silencing of a target gene by the Drosophila polycomb response element (PRE) but allow trans interactions between PREs on different chromosomes. Genetics 1997; 147:209-221.
86. Taillebourg E, Dura JM. A novel mechanism for P element homing in Drosophila. Proc Natl Acad Sci USA 1999; 96:6856-6861.

87. Buchenau P, Hodgson J, Strutt H et al. The distribution of polycomb-group proteins during cell division and development in Drosophila embryos: impact on models for silencing. J Cell Biol 1998; 141:469-481.
88. Wang L, Brown JL, Cao R et al. Hierarchical recruitment of polycomb group silencing complexes. Mol Cell 2004; 14:637-646.
89. Cao R, Zhang Y. SUZ12 is required for both the histone methyltransferase activity and the silencing function of the EED-EZH2 complex. Mol Cell 2004; 15:57-67.
90. Caretti G, Di Padova M, Micales B et al. The Polycomb Ezh2 methyltransferase regulates muscle gene expression and skeletal muscle differentiation. Genes Dev 2004; 18:2627-2638.
91. Tolhuis B, Muijrers I, de Wit E et al. Genome-wide profiling of PRC1 and PRC2 Polycomb chromatin binding in Drosophila melanogaster. Nat Genet 2006; 38:694-699.
92. Ringrose L, Paro R. Epigenetic regulation of cellular memory by the Polycomb and Trithorax group proteins. Annu Rev Genet 2004; 38:413-443.
93. Poux S, McCabe D, Pirrotta V. Recruitment of components of Polycomb Group chromatin complexes in Drosophila. Development 2001; 128:75-85.
94. Mulholland NM, King IF, Kingston RE. Regulation of Polycomb group complexes by the sequence-specific DNA binding proteins Zeste and GAGA. Genes Dev 2003; 17:2741-2746.
95. Americo J, Whiteley M, Brown JL et al. A complex array of DNA-binding proteins required for pairing-sensitive silencing by a polycomb group response element from the Drosophila engrailed gene. Genetics 2002; 160:1561-1571.
96. Kozma G, Bender W, Sipos L. Replacement of a Drosophila Polycomb response element core and in situ analysis of its DNA motifs. Mol Genet Genomics 2008; 279:595-603.
97. Boyer LA, Plath K, Zeitlinger J et al. Polycomb complexes repress developmental regulators in murine embryonic stem cells. Nature 2006; 441:349-353.
98. Kuzmichev A, Nishioka K, Erdjument-Bromage H et al. Histone methyltransferase activity associated with a human multiprotein complex containing the Enhancer of Zeste protein. Genes Dev 2002; 16:2893-2905.
99. Bernstein E, Duncan EM, Masui O et al. Mouse polycomb proteins bind differentially to methylated histone H3 and RNA and are enriched in facultative heterochromatin. Mol Cell Biol 2006; 26:2560-2569.
100. Hagstrom K, Muller M, Schedl P. A Polycomb and GAGA dependent silencer adjoins the Fab-7 boundary in the Drosophila bithorax complex. Genetics 1997; 146(4):1365-1380.
101. Hodgson JW, Argiropoulos B, Brock HW. Site-specific recognition of a 70-base-pair element containing d(GA)(n) repeats mediates bithoraxoid polycomb group response element-dependent silencing. Mol Cell Biol 2001; 21:4528-4543.
102. Schweinsberg S, Hagstrom K, Gohl D et al. The enhancer-blocking activity of the Fab-7 boundary from the Drosophila bithorax complex requires GAGA-factor-binding sites. Genetics 2004; 168:1371-1384.
103. Beisel C, Buness A, Roustan-Espinosa IM et al. Comparing active and repressed expression states of genes controlled by the Polycomb/Trithorax group proteins. Proc Natl Acad Sci USA 2007; 104:16615-16620.
104. Blastyak A, Mishra RK, Karch F et al. Efficient and specific targeting of Polycomb group proteins requires cooperative interaction between Grainyhead and Pleiohomeotic. Mol Cell Biol 2006; 26:1434-1444.
105. Brown JL, Grau DJ, DeVido SK et al. An Sp1/KLF binding site is important for the activity of a Polycomb group response element from the Drosophila engrailed gene. Nucleic Acids Res 2005; 33:5181-5189.
106. Biggin MD, Tjian R. Transcription factors that activate the Ultrabithorax promoter in developmentally staged extracts. Cell 1988; 53:699-711.
107. Mohd-Sarip A, Venturini F, Chalkley GE et al. Pleiohomeotic can link polycomb to DNA and mediate transcriptional repression. Mol Cell Biol 2002; 22:7473-7483.
108. Kal AJ, Mahmoudi T, Zak NB et al. The Drosophila brahma complex is an essential coactivator for the trithorax group protein zeste. Genes Dev 2000; 14:1058-1071.
109. Saurin AJ, Shao Z, Erdjument-Bromage H et al. A Drosophila Polycomb group complex includes Zeste and dTAFII proteins. Nature 2001; 412:655-660.
110. Dejardin J, Cavalli G. Chromatin inheritance upon Zeste-mediated Brahma recruitment at a minimal cellular memory module. EMBO J 2004; 23:857-868.
111. Hur MW, Laney JD, Jeon SH et al. Zeste maintains repression of Ubx transgenes: support for a new model of Polycomb repression. Development 2002; 129:1339-1343.
112. Wang YJ, Brock HW. Polyhomeotic stably associates with molecular chaperones Hsc4 and Droj2 in Drosophila Kc1 cells. Dev Biol 2003; 262:350-360.

113. Dellino GI, Schwartz YB, Farkas G et al. Polycomb silencing blocks transcription initiation. Mol Cell 2004; 13:887-893.
114. Reinberg D, Sims RJ 3rd. de FACTo nucleosome dynamics. J Biol Chem 2006; 281:23297-23301.
115. Zhou W, Zhu P, Wang J et al. Histone H2A monoubiquitination represses transcription by inhibiting RNA polymerase II transcriptional elongation. Mol Cell 2008; 29:69-80.
116. Stock JK, Giadrossi S, Casanova M et al. Ring1-mediated ubiquitination of H2A restrains poised RNA polymerase II at bivalent genes in mouse ES cells. Nat Cell Biol 2007; 9:1428-1435.
117. Kouzarides T. Chromatin modifications and their function. Cell 2007; 128:693-705.
118. Arrigoni R, Alam SL, Wamstad JA et al. The Polycomb-associated protein Rybp is a ubiquitin binding protein. FEBS Lett 2006; 580:6233-6241.
119. Bejarano F, Gonzalez I, Vidal M et al. The Drosophila RYBP gene functions as a Polycomb-dependent transcriptional repressor. Mech Dev 2005; 122:1118-1129.
120. Trimarchi JM, Fairchild B, Wen J et al. The E2F6 transcription factor is a component of the mammalian Bmi1-containing polycomb complex. Proc Natl Acad Sci USA 2001; 98:1519-1524.
121. Courel M, Friesenhahn L, Lees JA. E2f6 and Bmi1 cooperate in axial skeletal development. Dev Dyn 2008; 237:1232-1242.
122. Ogawa H, Ishiguro K, Gaubatz S et al. A complex with chromatin modifiers that occupies E2F- and Myc-responsive genes in G0 cells. Science 2002; 296:1132-1136.
123. Attwooll C, Oddi S, Cartwright P et al. A novel repressive E2F6 complex containing the polycomb group protein, EPC1, that interacts with EZH2 in a proliferation-specific manner. J Biol Chem 2005; 280:1199-1208.
124. Petruk S, Sedkov Y, Riley KM et al. Transcription of bxd noncoding RNAs promoted by trithorax represses Ubx in cis by transcriptional interference. Cell 2006; 127:1209-1221.
125. Smith ST, Petruk S, Sedkov Y et al. Modulation of heat shock gene expression by the TAC1 chromatin-modifying complex. Nat Cell Biol 2004; 6:162-167.
126. Eissenberg JC, Shilatifard A, Dorokhov N et al. Cdk9 is an essential kinase in Drosophila that is required for heat shock gene expression, histone methylation and elongation factor recruitment. Mol Genet Genomics 2007; 277:101-114.
127. Gregory GD, Vakoc CR, Rozovskaia T et al. Mammalian ASH1L is a histone methyltransferase that occupies the transcribed region of active genes. Mol Cell Biol 2007; 27:8466-8479.
128. Klymenko T, Muller J. The histone methyltransferases Trithorax and Ash1 prevent transcriptional silencing by Polycomb group proteins. EMBO Rep 2004; 5:373-377.
129. Srinivasan S, Armstrong JA, Deuring R et al. The Drosophila trithorax group protein Kismet facilitates an early step in transcriptional elongation by RNA Polymerase II. Development 2005; 132:1623-1635.
130. Ali JY, Bender W. Cross-regulation among the polycomb group genes in Drosophila melanogaster. Mol Cell Biol 2004; 24:7737-7747.
131. Cho YS, Kim EJ, Park UH et al. Additional sex comb-like 1 (ASXL1), in cooperation with SRC-1, acts as a ligand-dependent coactivator for retinoic acid receptor. J Biol Chem 2006; 281:17588-17598.
132. Kumar S, Jayaraman K, Panchanathan S et al. BEST: a novel computational approach for comparing gene expression patterns from early stages of Drosophila melanogaster development. Genetics 2002; 162:2037-2047.
133. Ringrose L, Ehret H, Paro R. Distinct contributions of histone H3 lysine 9 and 27 methylation to locus-specific stability of polycomb complexes. Mol Cell 2004; 16:641-653.
134. Bird A. DNA methylation patterns and epigenetic memory. Genes Dev 2002; 16:6-21.
135. Vire E, Brenner C, Deplus R et al. The Polycomb group protein EZH2 directly controls DNA methylation. Nature 2006; 439:871-874.
136. Hernandez-Munoz I, Lund AH, van der Stoop P et al. Stable X chromosome inactivation involves the PRC1 Polycomb complex and requires histone MACROH2A1 and the CULLIN3/SPOP ubiquitin E3 ligase. Proc Natl Acad Sci USA 2005; 102:7635-7640.
137. Mohn F, Weber M, Rebhan M et al. Lineage-specific polycomb targets and de novo DNA methylation define restriction and potential of neuronal progenitors. Mol Cell 2008; 30:755-766.
138. Negishi M, Saraya A, Miyagi S et al. Bmi1 cooperates with Dnmt1-associated protein 1 in gene silencing. Biochem Biophys Res Commun 2007; 353:992-998.
139. Rountree MR, Bachman KE, Baylin SB. DNMT1 binds HDAC2 and a new corepressor, DMAP1, to form a complex at replication foci. Nat Genet 2000; 25:269-277.
140. Lyko F, Ramsahoye BH, Jaenisch R. DNA methylation in Drosophila melanogaster. Nature 2000; 408:538-540.
141. Ferres-Marco D, Gutierrez-Garcia I, Vallejo DM et al. Epigenetic silencers and Notch collaborate to promote malignant tumours by Rb silencing. Nature 2006; 439:430-436.
142. Schwendemann A, Lehmann M. Pipsqueak and GAGA factor act in concert as partners at homeotic and many other loci. Proc Natl Acad Sci USA 2002; 99:12883-12888.

143. Gullerova M, Proudfoot NJ. Cohesin complex promotes transcriptional termination between convergent genes in S. pombe. Cell 2008; 132:983-995.
144. Hake SB, Allis CD. Histone H3 variants and their potential role in indexing mammalian genomes: the "H3 barcode hypothesis". Proc Natl Acad Sci USA 2006; 103:6428-6435.
145. Hake SB, Garcia BA, Duncan EM et al. Expression patterns and posttranslational modifications associated with mammalian histone H3 variants. J Biol Chem 2006; 281:559-568.
146. Janicki SM, Tsukamoto T, Salghetti SE et al. From silencing to gene expression: real-time analysis in single cells. Cell 2004; 116:683-698.
147. Zhang Z, Shibahara K, Stillman B. PCNA connects DNA replication to epigenetic inheritance in yeast. Nature 2000; 408(6809):221-225.
148. Tagami H, Ray-Gallet D, Almouzni G et al. Histone H3.1 and H3.3 complexes mediate nucleosome assembly pathways dependent or independent of DNA synthesis. Cell 2004; 116:51-61.
149. Ng RK, Gurdon JB. Epigenetic inheritance of cell differentiation status. Cell Cycle 2008; 7(9):1173-1177.
150. Ng RK, Gurdon JB. Epigenetic memory of an active gene state depends on histone H3.3 incorporation into chromatin in the absence of transcription. Nat Cell Biol 2008; 10:102-109.
151. Yamamoto Y, Girard F, Bello B et al. The cramped gene of Drosophila is a member of the Polycomb-group and interacts with mus209, the gene encoding Proliferating Cell Nuclear Antigen. Development 1997; 124:3385-3394.
152. Fanti L, Perrini B, Piacentini L et al. The trithorax group and Pc group proteins are differentially involved in heterochromatin formation in Drosophila. Chromosoma 2008; 117:25-39.
153. Fang J, Chen T, Chadwick B et al. Ring1b-mediated H2A ubiquitination associates with inactive X chromosomes and is involved in initiation of X inactivation. J Biol Chem 2004; 279:52812-52815.
154. Mak W, Baxter J, Silva J et al. Mitotically stable association of polycomb group proteins eed and enx1 with the inactive x chromosome in trophoblast stem cells. Curr Biol 2002; 12:1016-1020.
155. Pasini D, Hansen KH, Christensen J et al. Coordinated regulation of transcriptional repression by the RBP2 H3K4 demethylase and Polycomb-Repressive Complex 2. Genes Dev 2008; 22:1345-1355.
156. Ficz G, Heintzmann R, Arndt-Jovin DJ. Polycomb group protein complexes exchange rapidly in living Drosophila. Development 2005; 132:3963-3976.
157. Lee TI, Jenner RG, Boyer LA et al. Control of developmental regulators by Polycomb in human embryonic stem cells. Cell 2006; 125:301-313.
158. Sustar A, Schubiger G. A transient cell cycle shift in Drosophila imaginal disc cells precedes multipotency. Cell 2005; 120:383-393.
159. McConnell SK, Kaznowski CE. Cell cycle dependence of laminar determination in developing neocortex. Science 1991; 254:282-285.
160. Ambros V. Cell cycle-dependent sequencing of cell fate decisions in Caenorhabditis elegans vulva precursor cells. Development 1999; 126:1947-1956.
161. Ohsugi K, Gardiner DM, Bryant SV. Cell cycle length affects gene expression and pattern formation in limbs. Dev Biol 1997; 189:13-21.
162. Klebes A, Sustar A, Kechris K et al. Regulation of cellular plasticity in Drosophila imaginal disc cells by the Polycomb group, trithorax group and lama genes. Development 2005; 132:3753-3765.
163. Lee N, Maurange C, Ringrose L et al. Suppression of Polycomb group proteins by JNK signalling induces transdetermination in Drosophila imaginal discs. Nature 2005; 438:234-237.
164. McClure KD, Schubiger G. A screen for genes that function in leg disc regeneration in Drosophila melanogaster. Mech Dev 2008; 125:67-80.
165. Gregoire D, Brodolin K, Mechali M. HoxB domain induction silences DNA replication origins in the locus and specifies a single origin at its boundary. EMBO Rep 2006; 7:812-816.
166. Norio P. DNA replication: the unbearable lightness of origins. EMBO Rep 2006; 7:779-781.
167. Fisher D, Mechali M. Vertebrate HoxB gene expression requires DNA replication. EMBO J 2003; 22:3737-3748.
168. Hartl T, Boswell C, Orr-Weaver TL et al. Developmentally regulated histone modifications in Drosophila follicle cells: initiation of gene amplification is associated with histone H3 and H4 hyperacetylation and H1 phosphorylation. Chromosoma 2007; 116:197-214.
169. Luo L, Yang X, Takihara Y et al. The cell-cycle regulator geminin inhibits Hox function through direct and polycomb-mediated interactions. Nature 2004; 427:749-753.
170. Wohlschlegel JA, Dwyer BT, Dhar SK et al. Inhibition of eukaryotic DNA replication by geminin binding to Cdt1. Science 2000; 290:2309-2312.
171. Ohtsubo M, Yasunaga S, Ohno Y et al. Polycomb-group complex 1 acts as an E3 ubiquitin ligase for Geminin to sustain hematopoietic stem cell activity. Proc Natl Acad Sci USA 2008; 105:10396-10401.

172. Seo S, Herr A, Lim JW et al. Geminin regulates neuronal differentiation by antagonizing Brg1 activity. Genes Dev 2005; 19:1723-1734.
173. Park IK, Qian D, Kiel M et al. Bmi-1 is required for maintenance of adult self-renewing haematopoietic stem cells. Nature 2003; 423:302-305.
174. Lessard J, Sauvageau G. Bmi-1 determines the proliferative capacity of normal and leukaemic stem cells. Nature 2003; 423:255-260.
175. Dimri GP, Martinez JL, Jacobs JJ et al. The Bmi-1 oncogene induces telomerase activity and immortalizes human mammary epithelial cells. Cancer Res 2002; 62:4736-4745.
176. Molofsky AV, Pardal R, Iwashita T et al. Bmi-1 dependence distinguishes neural stem cell self-renewal from progenitor proliferation. Nature 2003; 425:962-967.
177. Rajasekhar VK, Begemann M. Concise review: roles of polycomb group proteins in development and disease: a stem cell perspective. Stem Cells 2007; 25:2498-2510.
178. Bernstein BE, Mikkelsen TS, Xie X et al. A bivalent chromatin structure marks key developmental genes in embryonic stem cells. Cell 2006; 125:315-326.
179. Azuara V, Perry P, Sauer S et al. Chromatin signatures of pluripotent cell lines. Nat Cell Biol 2006; 8:532-538.
180. Mikkelsen TS, Ku M, Jaffe DB et al. Genome-wide maps of chromatin state in pluripotent and lineage-committed cells. Nature 2007; 448:553-560.
181. Roh TY, Cuddapah S, Cui K et al. The genomic landscape of histone modifications in human T-cells. Proc Natl Acad Sci USA 2006; 103:15782-15787.

CHAPTER 4

Control of Vertebrate Hox Clusters by Remote and Global *Cis*-Acting Regulatory Sequences

François Spitz*

Abstract

Despite apparently shared structural organisation and functional roles, vertebrate Hox genes are controlled by regulatory mechanisms rather distinct from those of the prototypic *Drosophila* Antennapedia (ANT-C) and Bithorax (BX-C) Complexes. If individual regulatory modules have been shown to recapitulate specific Hox expression patterns, other experimental studies underscore that vertebrate Hox clusters are controlled in many of their functions in a global manner, through distinct mechanisms. We will discuss the different models that have been proposed to account for these global regulatory modes. In this context, the studies of the regulation of the HoxD complex during limb development highlighted the role of global regulatory elements and the different mechanisms associated to transform a structural organisation into distinct temporal and spatial expression domains. We will further discuss how these mechanisms may have benefited from the structure of the vertebrate homeotic clusters and reciprocally contribute to shape their evolution towards an increased level of organisation and compaction.

Introduction

The organization of developing structures along the metazoan antero-posterior axis requires the contribution of multiple gene products. An important step in embryogenesis is to pattern the developing embryo and establish regional identities that would be translated later in morphological differences. Amongst the multiple genes involved in this process, members of the Hox gene family have arisen as key players, not only because of the important roles they endorse during embryogenesis of various animals, but because their peculiar chromosomal organization and regulatory modes have captured the imagination of many scientists and because their studies have opened the way to many concepts that have been—more or less successfully—exported to other gene models.

In the fruit fly, early studies identified a large set of "homeotic" mutations, i.e., mutations that led to the transformation of a structure (e.g., an antenna) into another one (e.g., a leg). Thirty years ago, Ed Lewis showed that the genes associated with these homeotic transformations were clustered on the same chromosome and that their relative positions was colinear with the thoracic or abdominal segments which identity was under their respective control.[1] Later works showed that these genes encoded structurally related proteins, characterized by the "homeobox", a highly conserved DNA-binding domain.[2,3] In the 80s, several groups identified in different vertebrates—mostly human and mouse—genes encoding proteins with strong sequence similarities with the fly Hox genes.[4,5] Further works showed that, not only the proteins were similar, but that like in

*François Spitz—Developmental Biology Unit, European Molecular Biology Laboratory, Meyerhofstrasse 1, 69117 Heidelberg, Germany. Email: spitz@embl.de

Hox Genes: Studies from the 20th to the 21st Century, edited by Jean S. Deutsch.

Drosophila melanogaster, the vertebrate Hox genes were also organized in clusters and that their expression domains during mouse embryogenesis were colinear both in space and time to their position within these clusters.[6-8] The observed changes of vertebra morphology after gain- and loss-of-function experiments demonstrated further that vertebrate Hox genes controlled position identity along the antero-posterior axis of animal bodies and thus were analogue to their fly counterparts,[9] with little—if any—role in head structures.

These findings led to the view that a blueprint of animal body plans is structurally inscribed on their chromosomes by the Hox clusters and that the mechanisms that controlled their expression were responsible for a great part of the evolution of animal body plans. However, beyond this similarity between Hox genes organisation and expression specificities across different phyla, the nature of the vertebrate and fly Hox complexes, as well as their regulation are very different from one another. Whereas the *Drosophila melanogaster* homeotic genes are controlled separately through extended intergenic regions, the vertebrate clusters display an extremely compact organization as well as a very different mode of regulation. Their expression is also controlled by some intergenic *cis*-regulatory elements, but recent studies have underlined that important expression domains of the vertebrate homeotic genes are controlled by *cis*-acting sequences localised at a distance of the homeotic cluster, that act globally on multiple Hox genes within a complex.[10] This is notably the case for expression domains that have been gained by Hox clusters together with the emergence of new functions during vertebrate evolution. In this chapter, we will review the current state of knowledge on the regulatory mechanisms that control vertebrate gene expression and focus on the role of long range and global *cis*-acting elements, both for new expression domains such as the limb and for more ancestral features such as antero-posterior patterning. We will discuss how these mechanisms may have arisen and contributed to maintain Hox clustering and colinearity.

Colinearity and Clustering of the Homeotic Genes: An Obligatory Functional Link?

In many metazoan phyla, despite extensive variations in gene content (due to independent gene deletions or duplications) the Hox genes are essentially organised in clusters, formed by a succession of genes ordered according to their group of paralogy[9] (Fig. 1). This extreme conservation is not only a remain—and a proof of existence—of an ancestral bilaterian Hox cluster, but it is also highly suggestive that this precise structural organisation is an essential functional part of the mechanisms that control Hox gene expression. Hence, clustering and colinearity were considered as functionally interdependent through a universal mechanism that linked the structural organisation of Hox genes and their activities along the antero-posterior (A-P) axis.[11]

However, several recent works on different species have consistently challenged this view. In several animals, Hox clusters have undergone dramatic rearrangements leading to extremely dispersed and broken clusters, culminating with the "atomised" situation present in the urochordate *Oikopleura*.[12-14] Strikingly, in these different species, the now lonely Hox genes are still expressed differentially along the developing A-P axis, showing some sort of pseudo-colinearity between their expression profile and their ancestral—but lost—structural organisation. These observations suggested that the peculiar clustered and highly ordered organization of the Hox genes could be dispensable for their proper expression. In fact, several transgenic experiments had previously raised similar issues by showing that single vertebrate Hox genes integrated randomly in the mouse genome could still fairly reproduce the expression profile of their endogenous counterpart (see for example,[15-19]) putting into question the need for clustering. However, in these cases, several aspects of Hox gene expression were not recapitulated. Often, these Hox transgenes were lacking some expression domains of the endogenous genes, or their expression boundaries were slightly shifted along the A-P axis. The same remarks could be addressed to the "atomised" homeotic complexes found in *Oikopleura* and others: if the genes maintained a sort of spatial colinearity, it was not the case for the temporal colinearity which has been observed in vertebrates, but also reported in cephalochordates (*Amphioxus*),[20] insects (*Tribolium*)[21] and lophotrochozoans (*Chaetopterus*).[22]

Therefore, one can argue that animals with atomised Hox complexes may have evolve a mode of development which may tolerate a relaxed coordination of Hox gene expression, enabling a dispersal of the genes and a loss of the ancestral mechanism associated with structural colinearity.[23] There is indeed little doubts that ancestral Hox genes were organised as a large cluster. However, it is less clear whether the regulatory mechanisms that controlled their activities have been preserved in extant animals. Hox genes and Hox clusters may be controlled by completely different means in different phyla, with their own utilisation and constraint on the structural nature of the clusters.[23,24]

Vertebrate Hox Clusters Are More Clustered Than Others

Beyond a common qualification and a roughly similar organisation, Hox clusters from different animals have very distinct features. Some "clusters" have undergone extensive rearrangements, to the point that this characteristic can barely be used to describe them and even in animals that have maintained a strict clustering of these genes, the size of a Hox cluster can vary considerably. For example, the overall size of the *Tribolium castaneum* Hox cluster is about 710 kb for a total genome size of 160-204 Mb,[21] whereas a typical mouse Hox cluster covers around 100 kb for a 2,700 Mb genome (Fig. 1). This would correspond to a ~100-fold compaction since gene and genome sizes are generally proportional. The vertebrate Hox clusters appear therefore as the upmost cases of clustering and compaction and seem to be oddities compared to other animals. It is unlikely that such a situation corresponded to the ancestral one and therefore it should have evolved in the vertebrate lineage by a process of consolidation.[23] The emergence of global or pan-cluster regulatory systems acting in multiple genes rather than on single ones is usually proposed to be a major factor which may have promoted such a phenomenon, paralleling the evolution of new functions for Hox genes in vertebrates. As a result of these processes, the clustered nature of the vertebrate Hox genes has been drastically re-enforced, while their regulation has also evolved towards more coordination. Different mechanisms can account for these coordinated and tissue- or organ-specific regulations, from shared *cis*-regulatory elements to large scale epigenetic and chromatin remodelling mechanisms. Indeed, a number of extensive analyses of Hox gene expression have shown that a large variety of these different strategies are contributing to vertebrate Hox gene regulation.

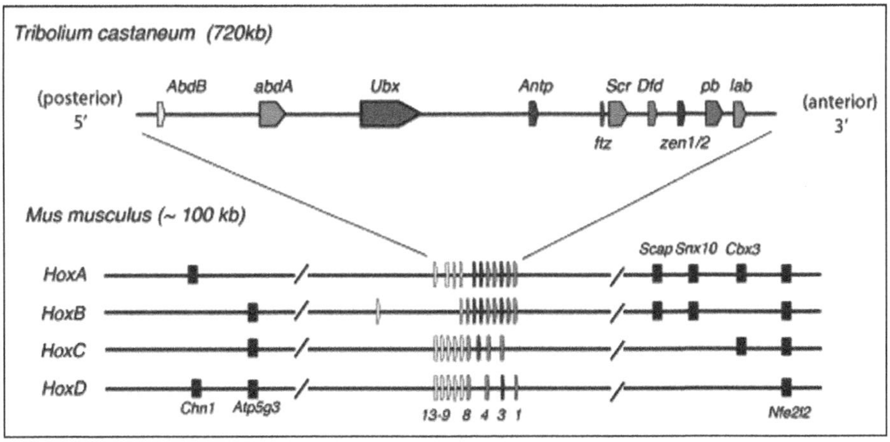

Figure 1. Homeotic clusters in insects and mammals. To scale representation of the *Tribolium castaneum* and mouse homeotic clusters. The vertebrate Hox clusters are very compact compared to insect clusters, but they are embedded within larger regions showing extensive conservation of gene order, as shown by the presence of conserved paralogous genes (for thanks of simplification, only few examples of these genes have been shown, not to-scale).

Global Regulation of the Complex through Shared Mechanisms: The Retinoic Acid Connection

The possibilities of a shared mechanism to control Hox genes have been proposed since the discovery of their clustered nature. In his landmark article,[1] Ed Lewis predicted that the *Drosophila* Hox genes should be controlled, in addition to gene-specific elements, by a double gradient of repressor activity in the embryo defined (1) by a DNA-binding repressor showing a A-P-gradient of concentration (2) by *cis*-acting elements having progressively less affinity for the repressor as they are localised away from the most anterior genes. Quite exquisitely, the regulation of the vertebrate Hox genes obeys to such a model, with retinoic acid (RA) acting as a graded signal (even though it is an activating and not a repressing signal).[25] The anterior boundaries of Hox expression domains in the hindbrain are controlled by the concentration of retinoids[26,27] and by the presence of functional RA response elements (RAREs) in the vicinity of multiple Hox genes.[28-31] This sensitivity to RA is also colinear and moves along the complex as development proceeds, with genes having different windows of competence to respond to RA depending on their position in the complex. The specific spatial response of the different Hox genes depends on the nature of the different RAREs as shown by swapping experiments.[32,33] Interestingly, only the 3′ most Hox genes can be precociously activated by RA and exclusively when lying within their normal position in the 3′ most region of a complex.[34,35] It suggests that besides the intrinsic properties of these RARE-containing elements, distinct higher-order rules may apply. RA has certainly a role in Hox regulation and there are indications of an ancestral association with Hox clusters,[36,10] but the elegant solution of a gradient of molecules acting on binding sites of variable affinity cannot account on its own for all features of Hox colinearity.

High-Order Structures Over the Complex and Colinearity

The colinear activation of the Hox genes has also been proposed to correspond to a progressive transition from an inactive to an active chromatin structure, with genes being progressively made available for transcription in a 3′ to 5′ wave.[11,37] Such a model was supported by several experiments where an anterior gene was transplanted at a more posterior position and shown to adopt a posterior-type expression.[38,39] Also, a close analysis of the initial steps of Hox gene activation in the primitive streak suggested the existence of a process anticipating effective Hox gene expression and that the associated "permissiveness" is also established sequentially, from 3′ to 5′.[40] In further direct support for this model, Wendy Bickmore and colleagues showed that the sequential activation of the *Hoxb* genes during ES cell differentiation and mouse gastrulation is paralleled by a visible decondensation of the locus and a progressive looping of the locus starting at the 3′ end out from the chromosomal territory (CT), both of which being considered as signs of transcriptional activation.[41,42]

Such observations provide an elegant mechanistic explanation to the orchestrated activation of the homeotic clusters. However, this model cannot account for all the aspects of Hox gene expression. For example, no extrusion from the CT is observed during the activation of the HoxD cluster in the developing limbs, raising issues about how general such a mechanism can be.[43] A recent study published by Wendy Bickmore's group further highlighted the complexity of the relationship between nuclear positioning and transcriptional activation of a homeotic cluster or a gene.[44] They took advantage of a mouse line where a *Hoxb1* transgene has been inserted in the 5′ end of the HoxD complex, near the posterior *Hoxd13* gene. This transgene carries the regulatory elements that are sufficient to mirror the endogenous *Hoxb1* expression in the primitive streak and in hindbrain rhombomere 4 (r4) when inserted at ectopic positions in the genome. When inserted next to *Hoxd13*, the *Hoxb1-LacZ* transgene conserved its early expression in the primitive streak and even lead to an ectopic activation of *Hoxd13* here.[35] However, these activities were only transient and no expression was ever scored in r4. Interestingly, the *Hoxb1-LacZ* transgene was able to induce significant chromatin reorganisation (looping out of the CT and decondensation), both in the primitive streak and in r4, compared to other rhombomeres, albeit not to the levels observed for the endogenous gene.[44] This result showed that the transgene comprised sufficient information to carry out the chromatin and nuclear positioning changes associated with activity of the endogenous gene. The situation in r4 suggested that these changes act upstream transcriptional activation, since they occurred despite lack of detectable expression of the transgene in r4. However, one should point out that the changes induced by the transgene on chromatin organisation were only local; in terms of transcriptional activation, they were not sufficient to disrupt completely the colinear regulation of the complex, as the transgene was repressed in anterior regions—r4—and only transiently expressed in the primitive streak. Later on, the transgene fell under the control of the posterior *Hoxd* genes machinery and adopted the expression specificities of *Hoxd13*, its new neighbour.[35]

These experiments clearly underline that a global and colinear remodelling of chromatin over the whole complex is taking an important part in its regulation. It is easy to understand how such a mechanism can allow for a coordinated colinear activation of the complex and act as a constraint to maintain clustering.[24] However, these experiments highlighted also the recurrent dichotomy found in Hox gene regulatory controls, with local elements that seem to be sufficient for many of their specificities and global controls that can override them.

Control of Vertebrate Hox Genes by Shared Internal Enhancers

As mentioned previously and as exemplified by many transgenic analyses of Hox gene regulation, the proximal flanking regions of each Hox gene seem to contain regulatory elements that could autonomously reproduce—more or less accurately—the expression pattern of the corresponding endogenous gene, in response to different signals and transcription factors, such as RA, FGFs, WNTs and Cdxs. However, these experiments were done with single gene-transgenes, an approach that only offers a reductionist representation of a real Hox cluster. Given the compact and clustered nature of the Hox complexes and the small size of the genes, in vertebrates, any *cis*-regulatory element associated to a specific Hox gene would not be very far from the immediately adjacent Hox gene and thus, could impinge on its expression pattern as well. Indeed, experiments which were using a better representation of a Hox complex (either using multigenic transgenes or recombining the endogenous locus) revealed widespread sharing of regulatory elements between neighbouring Hox genes.[45-47] An extensive sharing of enhancers by Hox genes could result in chain-like system where successively intermingled regulatory interactions would maintain the overall cluster.[11,48] Importantly, enhancer sharing in the Hox clusters is also associated with mechanisms such as promoter competition and selectivity, which together contribute to control and refine the diverse transcriptional outputs provided by these elements.[45]

The complexity of these interactions between local enhancers and flanking genes suggests that reducing their contribution to a mere addition of activities, restricted to one or two target genes, may not be appropriate in a situation where all elements are so densely clustered and intermingled. Indeed, one can envision that these elements, having some autonomous activity, may function altogether as a pan-Hox cluster enhanceosome, which global activity would exceed the sum of the contribution of its individual elements and form a fully integrated functional and structural unit. In this view, the coherent regulation of Hox genes as an organised group may result from the complex network of interactions taking place between the multiple genes and *cis*-regulatory elements embedded within the clusters. Such a situation may offer multiple opportunities for evolutionary changes in the details of the role of each element. And thus, behind an apparent similarity in the control of Hox genes between different clusters and species, the associated regulatory "logic"—if any—may therefore be hard to "crack".

The Ins and Outs of *Hoxd* Gene Regulation

The *cis*-regulatory elements that are located within each Hox complex have undoubtedly important roles to play in their activation. However, as shown by the study of the mouse HoxD complex, sequences localised outside of the complex are as well required. In addition to patterning the vertebral column and neural tube, genes of the HoxD complex have also important roles in morphogenesis of the tetrapod appendages, external genitalia and digestive system (Fig. 2).[49,50] These functions are also endorsed by contiguous set of genes, in a manner that is reminiscent of the

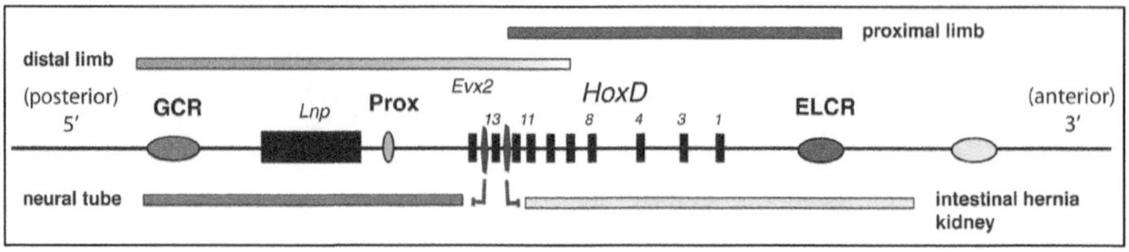

Figure 2. HoxD cluster, remote enhancers and regulatory landscapes. Distinct remote enhancers define overlapping regulatory landscapes (regions of shared expression patterns) over the *Lnp-Evx2*-HoxD locus. The different enhancers (identified—GCR, Prox[61,60]—or yet unlocalised—ELCR, gut[58] and kidney[59]) are represented by coloured ovals. Bars of the corresponding colours show the extent of the different landscapes (the left limit of the neural/distal limb landscapes are unknown). Some landscape boundaries are defined by the presence of boundary elements (e.g., neural landscape,[65,93] intestinal hernia[85]), which can be landscape-specific or not (shown as lozenges).

colinearity displayed in the main axis, suggesting also that these new features were controlled at the level of the cluster, rather than at individual gene level. In the limbs, colinearity appears in multiple aspects. In a first phase, *Hoxd* genes are activated throughout the whole emerging limb bud in a sequential manner starting with the most 3′ genes (*Hoxd1-d3*). As the limb grows, more 5′ genes are activated but become restricted to posterior nested domains. These domains are transformed into spatially distinct compartments along the limb proximal to distal axis and would correspond to the stylopod, zeugopod and autopod. Finally, a third type of colinearity is observed in the distal limb mesenchyme, where the most 5′ *Hoxd* (*Hoxd10-Hoxd13*) genes show an inverted colinearity along the A-P axis, with *Hoxd13* showing the stronger and more anterior expression.[51-53]

The observation that different transgenes (Hox and non-Hox) inserted into the posterior region of the HoxD complex adopted an expression pattern corresponding to their new position and not to their own regulatory elements suggested strongly that the complex was under a global regulation.[35,38,39] Interestingly, when a large BAC clone covering most of the human HOXD complex (from *HOXD3* to 30kb upstream *HOXD13)* was introduced in transgenic mice, no expression of the human HOXD genes were scored in the limbs or in the gut, whereas the correct expression domains were established in the trunk mesoderm and neural tube.[54] Conversely, the typical limb and intestinal *Hoxd* expression domains were still observed after the replacement of the 100 kb of the whole complex by single *Hoxd11-LacZ* transgene.[54] Altogether, these complementary experiments demonstrated that if sequences localised within the complex itself could elicit correct expression domains along the main body axis, other domains of expression of the *Hoxd* genes require "external" elements localised outside of the complex.

The Role of the Flanking Regions in the Control of Vertebrate Hox Genes

Other indirect evidences were suggesting more generally that the regions surrounding the Hox complexes could as well be involved in their regulation. Hox clusters are in the middle of large regions of conserved syntenies from fishes to mammals and this observation also extends to the paralogous clusters[55] (Fig. 1). The rather extensive conservation of the architecture of these loci could imply that the nature of the regions flanking the Hox clusters may be under selection pressure, possibly as they contain several critical regulatory elements. Indeed, these regions are densely filled with nongenic elements which sequences are conserved from fish to mammals.[55]

Mutations occurring in human and mouse have also suggested a global control of Hox genes through elements localised at a distance from the complex. For example, different chromosomal rearrangements occurring in the vicinity of the HOXD complex, either on the 3′ end or on the 5′ end, have been associated with malformations affecting regions of the body under Hox gene control.[56,57] A translocation with a breakpoint localised 60 kb downstream of the HOXD complex was particularly interesting since the affected patients displayed several vertebral anomalies (fusion of anterior cervical vertebrae, cleft of the lumbo-sacral ones) as well as shortened forearms, which can be interpreted as consecutive to a global mis-expression of the whole HOXD complex due to perturbation of the flanking sequences.[56]

Further works on the mouse model system, taking advantage of large scale transgenesis and chromosomal engineering, have helped to identify the corresponding elements, understand how they work on the different genes of the cluster and how they have contributed to maintain Hox genes as a tight cluster.

Control of the HoxD Cluster through Remote Enhancers

To understand whether clustering was indeed required for *Hoxd* genes function and to map the regulatory elements that were proposed to lie outside the complex, the group of Denis Duboule produced a large chromosomal inversion that split the complex between the *Hoxd11* and *Hoxd10* genes.[58] Importantly, in the wild-type situation, these two genes were both expressed in a proximal and a distal domain in the limb bud, in the genital and in the intestinal hernia, all domains dependent on the activity of "external" enhancers. Splitting the complex in two halves induce a partition in these expression patterns, with *Hoxd11* (and the 5′-half cluster) being still expressed in the autopod and genital bud, but not in the zeugopod and gut, whereas *Hoxd10* was showing the exactly opposite behaviour.[58] This was a direct demonstration that the overlapping patterns of *Hoxd* genes in these domains were defined by shared elements, localised on both sides of the complex (the 3′ end for early/proximal limb and gut; the 5′end for distal limb and genitalia). Additional enhancers controlling the expression of *Hoxd* genes in both the ureteric bud and metanephric mesenchyme of the developing kidney have also been mapped outside of the complex itself, on the 3′ end[59] (Fig. 2).

This specific arrangement, whereby external enhancers from both sides of the locus control the distinct but overlapping groups of genes, even if they correspond to relatively recent additions to the *Hoxd* regulatory portfolio, had likely contributed to keeping these genes as a tight cluster. In case of the distal limb expression domain, associated regulatory elements have been mapped to two distinct regions, the GCR (Global Control Region) and Prox, localised at 200 kb and 45 kb upstream *Hoxd13*, respectively.[60,61] Both regions can activate transcription from *Hoxd* genes or heterologous promoters in the distal limb buds. Their own activities are however distinct. Both are initiated in the posterior limbs, around day 10, at a time that coincides with the distal expression of the endogenous *Hoxd13*, but evolved distinct spatial specificities. The GCR-transgenes showed a posteriorly polarized crescent expression, similarly to *Hoxd13*, but which usually failed to extend up to the region of future digit I and with a rather low expression in cells from the medial part of the limb compared to the ones localised close to the ventral or dorsal ectoderm.[61] In contrast, Prox drove expression in the interdigital mesenchyme, essentially in the ventral region.[60] However, when the two elements are combined on a single transgene, they can activate *Hoxd11* reporter construct in a pattern reinforced throughout the mesenchyme and extended anteriorly and distally. A large transgene containing GCR and Prox upstream of the posterior half of the human HOXD complex can also correctly reproduce the late distal expression pattern of the genes and functionally complement—to some extent—the synpolydactyly caused by a deficiency for the murine *Hoxd13-11* genes.[60] These data show that the combination of GCR and Prox is sufficient to drive *Hoxd* gene expression in the distal limb bud.

Interestingly, this overall distal limb activity of the posterior *Hoxd* genes results from the synergistic combination of distinct enhancers with different but complementary specificities. The observation that even apparently homogeneous expression domains are not defined by a master regulatory element but rather by a large set of complementary and/or apparently redundant *cis*-regulatory elements seems to be a rather general feature of gene regulation.[62,63] It may correspond to a way to integrate different pathways into tissue- or organ-specific expression profiles, to refine the regulation of each gene and confer robustness to gene expression. It may also be a relic of the regulatory evolution of the locus, which may have proceeded through successive additions of elements, e.g., to build and pattern cartilage condensation of mesenchymal cells to stimulate their growth to form digits.

Importantly, in addition to distinct spatial activities, the Prox and GCR elements also seem to differ in important functional properties. Remarkably, in absence of GCR, Prox seemed to be unable to activate *Hoxd13* or *Hoxd11* when placed in its normal context, while it can do it when the reporter gene is juxtaposed just next to it. When GCR was not included in the previously used large human BAC transgene, *HOXD13* was not detected in the autopod and the construct did not rescue *Hoxd13* mutant synpolydactyly.[54] Similarly, the *Ulnaless* inversion, which separates the Prox-HoxD segment from GCR, leads to the disappearance of *Hoxd* gene expression in the distal limb bud.[54] This can be due to a limitation in the range of action of Prox (either in terms of physical distance, or number of genes) or to its inability to bypass boundary elements that can lie between Prox and the cluster to prevent its ectopic activation in the CNS. This inability is relieved in presence of GCR, suggesting that this element may play a central role in organising the regulation of the locus, not only through its own enhancer activity, but also by helping integrating the outputs of other enhancer elements spread over the region and extending their activities to a large region.

Regulation of the HoxD Cluster and More: Global Control Regions and Regulatory Landscapes

Besides its activity in limbs, GCR is able to drive gene expression also in different territories of the brain and in the neural tube.[61] Both limb and neural enhancers have been mapped to the same 2.5 kb core region of GCR and similar activities are displayed by human or chicken GCR.[60] These neural domains of expression do not correspond to regions where *Hoxd* genes are expressed during embryogenesis. However, they coincide with the expression patterns of *Evx2* and *Lnp*, two unrelated genes that lie between GCR and the HoxD complex. Quite surprisingly, both *Evx2* and *Lnp* are also expressed in the distal limb bud, in a pattern that closely mimicks, spatially and temporally, the activation of *Hoxd13* in the corresponding domains and depends on the presence of GCR (as shown by the absence of *Evx2* expression in *Ulnaless* mutant limbs).[61]

Thus, GCR behaves like a Global Control Region (hence its name), as it contributes to the regulation of multiple and unrelated genes (*Lnp*, *Evx2* and the 5′ *Hoxd* genes) spread over a large chromosomal domain (>200 kb) and in different cell types (limb mesenchyme, neural tube, brain etc.). Any gene inserted within the locus will fall under its influence and adopt the corresponding expression profile, as shown by relocation experiments of *Hoxd9-LacZ* or *PGK-neo* transgenes in the *Lnp-Evx2* interval.[39,64] Rather than interacting specifically with target promoter regions, GCR (together with Prox and possibly other regulatory elements) defines large domains—called regulatory landscapes—where all genes adopt similar expression profiles, on the top on their own specificities. The extent of these landscapes can be limited by sharp boundaries, as illustrated by the neural landscape, which is restricted to *Lnp* and *Evx2* by an insulator-like element localised between *Evx2* and *Hoxd13*.[65] It could also be limited by an upper limit to the number of genes that a landscape can accommodate (Fig. 2). Indeed, as shown by the analysis of deletions and duplications within the HoxD cluster, limb expression starts to decrease after the fourth or fifth gene from the 5′ end of the cluster, mostly irrespectively of the nature of its promoter. For example, when put in the fourth position, a copy of *Hoxd13* is only weakly expressed in the limb compared to when it is at its normal first position.[66] Insertions of genes upstream of the complex can also induce mild changes in the activities of the more downstream ones.[64]

In this respect, the topology of the locus appears to be an important determinant of the overall output, as the most 5′ *Hoxd* gene is always the one expressed at the highest level and in the more anterior region of the autopod, whereas the more 3′ ones display gradually reduced expression. These spatial and quantitative colinearities are therefore intimately linked to the structural organisation of the cluster, much like what is seen for the body axis. However, in the distal limb, the direction of the colinearity has been reversed[53] with a predominant and broader expression of the "posterior" 5′ genes over more 3′ ones. From a functional perspective however, *Hoxd13* seems to be the major gene involved in the specification of the autopod as loss of function mutations of the other genes leads only to minor or no autopod malformations, as long as a functional HOXD13 protein is present.[67-69] Therefore, it is unclear whether the expression of the other genes has any real function per se, which in the case of *Hoxd* genes would probably be buffered by *Hoxd13* "posterior prevalence".[70] It may be well possible that this global expression reflects only the inherent sloppiness of the mechanism that ensures contact between GCR/Prox and *Hoxd13*. Alternatively, these expressions of the neighbouring genes may participate to the regulation of the locus. They may titrate enhancer activity to finely tune the expression level of *Hoxd13* in the limbs. The molecular details of the mechanism "linking" GCR/Prox to *Hoxd13* are yet unknown, but the experimental observations are consistent with a looping system as proposed for the beta-globin LCR.[71] However, whether the establishment of this favoured loop requires some scanning or facilitated looping which would be helped by the previous or transient recognition—and activation—of the promoter regions of the *Lnp* and *Evx2* genes remains to be tested.

Importantly, a more quantitative analysis of the system showed that the efficiency of the interaction between GCR and the different genes in the locus is not purely determined by their relative proximity, but also by differential "affinities" between the different promoter regions and the remote enhancers.[72] Interestingly, this approach revealed also that the total amount of Hox genes expressed in the autopod is a linear function of the number of genes present. This suggests that the transcriptional output is not determined by the intrinsic potential of the "limb enhancer" but corresponds to a delicate equilibrium, established through the regulatory architecture of the locus as well as gene-specific features. In an evolutionary context, such a situation would provide enough regulatory flexibility to facilitate morphological adaptations of the limbs. It can also result from a compromise between the regulatory constraints associated with a pleiotropic regulatory mechanism, used by one set of genes in the brain and yet by another set in the limbs and in the genitals, and the need for distinct regulations to better suit the developmental program of each of these structures.

As said earlier, the expression of the *Hoxd* genes in the early limbs depends on a distinct regulatory element localised at the other end of the cluster. Here again, this element seems to act on a global manner and its effects on a given *Hoxd* gene are determined by relative position of the gene in the cluster. However, whereas late distal limb regulation relies mostly on the gene rank calculated from the 5′ end, the expression in the early limb is defined by two opposite regulatory influences.[73] Firstly, *Hoxd* genes are activated sequentially throughout the newly emerging limb bud, starting with the most 3′ gene. This distance-dependent temporal activation is somehow counteracted by a mechanism repressing genes in the anterior part of the limbs and which is more efficient as genes are closer to the 5′ end of the cluster (Fig. 3). The balance between these opposing influences led to the establishment in the growing limbs of spatially nested domains centred at the posterior distal end. The sequences involved in these different activities have not been identified yet and thus it is not known whether they are defined by classical regulatory elements (enhancer/silencer) or involve progressive modifications of the chromatin or nuclear positioning of the locus (as discussed previously). However, given the similarity between this process and the colinear activation of Hox genes along the main body axis, it is tempting to speculate that the associated regulatory systems may be tightly related and hence that the patterning of the early limb by Hox genes along the limb proximal-distal/posterior-anterior axes is relying on the toolbox used to define axial antero-posterior identity along the main body, maybe complemented by limb-specific elements. On the top of these, a distinct step involves the recruitment of a new regulatory system (GCR/Prox) to further develop the most distal region of the limbs, using a different logic but again intimately coupled to the structure of the locus (Fig. 3).

Remote Enhancers for the Other Vertebrate Hox Clusters?

These studies shed light on the important contribution of remote and global enhancers to the regulation of the HoxD cluster. Currently, we unfortunately lack similar functional analysis for the other clusters and therefore it is unclear how general are these mechanisms. Anecdotal reports of developmental malformations or blood disorders associated with rearrangements occurring close to other vertebrate Hox clusters suggest that remote sequences can impinge on their regulation,[74-76] but the corresponding elements are unknown. Comparison of the sequences within and flanking the paralogous vertebrate Hox clusters highlighted only few conserved elements.[55,77] Amongst

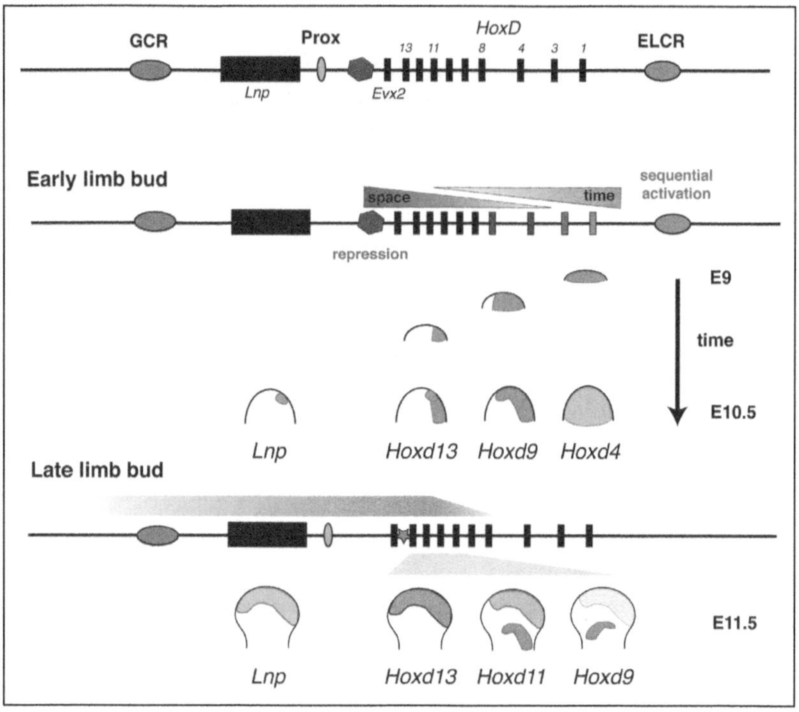

Figure 3. Structural and functional colinearities in the limb. The apparent colinearities of the *Hoxd* genes in the limb are the consequence of the independent regulations, which influences depend on the relative position of the gene within the cluster. As the limb emerges from the flank of the embryo, *Hoxd* genes are sequentially activated throughout the mesenchyme, by an Early Limb Control Region (ELCR), the genes closer to the ELCR (3′ end) being activated earlier. As the limb grows, these domains are progressively expanded along the proximo-distal axis, but, a negative element (hexagon) localised on the 5′ end of the cluster represses their extension into the anterior compartment of the limb, with a stronger repression for the genes closer to the 5′ end. The corresponding domains give rise to the zeugopod and stylopod domains at later stages of limb development. Overlapping with the end of the first phase, a second and independent activation takes place through the action of the GCR/Prox elements (dark and light grey) and turn genes on in the most distal part of the limb, forming a posteriorly polarized crescent. This crescent-like domains of expression comprises the future digit I for *Hoxd13* (and *Evx2* and *Lnp*), but are restricted more posteriorly for genes localised further away. The quantitative output of this regulation is also maximal for the *Hoxd* genes localised at the 5′ end of the cluster. Both these spatial and quantitative characteristics are associated with a putative tethering activity (star) localised tentatively between *Evx2* and *Hoxd13* (adapted from ref. 73).

these, two highly similar sequences are found upstream the HoxA and HoxD complexes, the first one next to *Evx1/Evx2* genes, the second one corresponding to the GCR.[78] The first element may be involved in the regulation of Evx genes,[79] but it also falls in a region that has been proposed to mediate global repression over the posterior end of the HoxD complex.[39] The HoxA-associated "GCR"-like element is localised 350kb from *Hoxa13* within an intron of the *Hibadh* gene. Intriguingly, *Hoxa13* and the four upstream genes spread over 1Mb are both expressed in the distal limbs and external genitalia[78] in a manner very reminiscent to the HoxD-*Evx2-Lnp* regulatory landscape. However, this activity is not directed by the HoxA-GCR-like element, even though it seems to act as an enhancer for a variety of other tissues.[78] This situation nevertheless suggests that some expression domains of the HoxA complex may be controlled by remote and promiscuous enhancers, similarly to GCR/Prox for HoxD and that an ancestral sequence with probably some kind of brain/neuronal enhancer activity was present at a distance from the complex before the duplication(s) that gave rise to these two Hox clusters. Yet, it is unclear how the ancestral GCR may have been involved in building up Hox gene regulation with remote enhancers, as the two derived elements have currently evolved different functional properties. But the pre-existence of a regulatory element involved already in long-range regulations (maybe with *Evx2* rather than *Lnp* or posterior *Hoxd* genes) may have fostered the emergence of new regulations, to create,[80] re-enforce or modify pre-existing limb expression domains.[81-83] Under this "regulatory priming" model,[60] ancestral regulatory elements could contribute either directly by providing a platform of transcription factors binding sites, which could evolve into a multifunctional enhancer with diverse specificities and target genes (e.g., GCR), or indirectly, by establishing an accessible chromosomal domain that would help genes to make use of rapidly evolving sequences to built new regulatory elements (such as Prox). Sequence comparison failed to reveal any trace of a putative ancestral GCR in amphioxus, even though both *metaxin2* and *lunapark* orthologs are flanking its homeotic cluster.[84]

An Evolutionary Success Story and an Increasing Need for a Global Regulation

Vertebrate Hox clusters have experienced a parallel evolution, with the acquisition of new functions and an increased level of structural organisation. As suggested by Denis Duboule, the "consolidation" of Hox clusters towards more compactness and organisation and their recruitment for new functions were likely mutually re-enforcing processes.[23] The clustered nature of the Hox genes played probably an important role in their evolutionary "success", as several genes with coherent functions could be recruited at once, providing more flexibility and evolutionary opportunities, with the Hox complex working as a meta-gene.[37] As this process was driven by regulatory sequences located outside of the complex, it has further re-enforced the need for clustering (to maintain the genes under the same control) and paved the way to develop new regulations (through "regulatory priming"). Several constraints were probably associated with these increasing functions and the need to accommodate all these different regulations. This may have led to the emergence of ad hoc solutions, such as "polar" silencers,[65,85] tethering elements[66] or even elimination of some Hox genes. These constraints may have been relaxed with the successive duplications of the complexes, which could have also allowed further co-options of the Hox genes into new important functions associated with major innovations in vertebrate body shape, such as "improved" appendages. The relative simplification of the teleost Hox complexes (7 complexes with 48 genes versus 4 complexes with 39 genes in total in mammals) could be interpreted as a step further in organizing Hox metagenes with increased functionalities.[23]

However, despite the diversities of the new functions acquired during evolution and as shown by Capecchi, Wellik and colleagues, the global patterning of the mammalian skeleton is defined by the overall contribution of paralogous genes from the four clusters.[86,87] There was therefore no partitioning of the ancestral function after cluster duplications, suggesting that either the mechanism of colinearity is deeply inscribed in the genes themselves, or—non-exclusively—that it was critical to maintain all Hox genes under strict control. Indeed, given the phenotypic dominance of Hox genes over more "anterior" ones—the so-called "posterior prevalence"—ectopic expression of Hox genes may have dramatic consequences (e.g.,[35,38]). The regulatory problem associated with this constraint may have become increasingly complex with the duplications of the ancestral Hox cluster and the need to coordinate these genes also in *trans* (between different complexes). Therefore, these duplications may have led to an increased need for coordinated regulation, which may have been difficult to achieve without the reinforcement of global mechanisms, acting on a whole cluster, on the top of individual solutions. This may have been possible by emphasizing the role of a pre-existing system responsible of the initiation of the different axial expression domains (temporal colinearity) and reinforcing it further and later during development through regulatory mechanisms acting in *trans*, involving cross-regulation between the clusters, both through HOX proteins[46,88,89] and a variety of miRNAs—some also ancestral—[90] and noncoding RNAs.[91]

Hox cluster duplications, by offering opportunities to evolve new functions and stressing pre-existing constraints, may have ineluctably triggered the evolution of each resulting cluster towards more compactness and global regulation. The regulatory mechanisms controlling Hox gene expression in vertebrates are the output of the evolution process which accommodated useful new opportunities with the regulatory constrains imposed by pre-existing Hox functions and need for tight regulation. The re-iterative regulatory tinkering that occurred at these loci led to a situation dominated by an elaborate hierarchy of *cis*-regulatory influences defined by intermingled elements both at the functional and structural levels, which may have considerably changed from those that were at play in the ancestral urbilaterian Hox complex. Such processes are certainly not specific to Hox genes and should have also contributed to remodel the regulatory architecture of other developmental genes. In the Hox case however, the striking feature is that these gene regulations take advantage of the structural organisation of the complex, with global enhancers acting in a position-dependent rather than in a strictly gene-specific manner. The underlying mechanisms can be distinct in their details, from progressive activation during time to differential quantitative efficiency, but they all manage to translate in the embryo the genomic architecture of the locus, illustrating probably that the "success" of these genes lies in their ability to act as a integrated patterning metagene.

Conclusion and Outlook for Hox Gene Regulation in the 21st Century

The last decade has seen major progresses in our understanding of Hox clusters regulation. These were driven mostly by extended analyses of the vertebrate loci by transgenesis and chromosomal engineering, which allowed the identification of many regulatory elements that control the transcriptional activities of these genes. In parallel, different studies of the conformation and chromatin structure of these loci in vivo have started to reveal other potential levels of regulation. The key challenge for the future will be to integrate these different strategies and conceptual frameworks to provide a more integrated view of Hox gene regulation. This integration is needed to define the hierarchy and the precise roles of the different elements in gene repression, activation and maintenance, as well as their relationship with the chromatin status and position of Hox loci within the cell nucleus. This will require (1) the development of novel genetic approaches to better probe the regulation of the locus in vivo and assess the role of individual elements (2) the adaptation of the protocols that have successfully described the chromatin status of the different mammalian complexes in cultured cells to in vivo situations. This latter point is currently a real challenge, firstly because of the limited number of cells available from an embryo, but also because the reduced ability of most current approaches to capture the dynamic changes that are likely to happen during activation of the Hox clusters. Given the recent and tremendous progress of microscopy, it is however not entirely unrealistic to envision future strategies enabling live imaging of the behaviour of Hox loci during mouse embryogenesis, as it is currently done in yeast.

We already know quite a number of elements controlling Hox gene expression, many being sufficient to recapitulate most of the expression specificities of these genes. However, survey of the large regions flanking these gene complexes have identified multiple additional elements with yet unknown activities, but which are extremely conserved at the sequence level and for many of them, displayed also biochemical features characteristic of gene regulatory elements.[92] Thus, the already large number of Hox gene regulatory elements may further increase in the near future, reflecting the extensive modularity and redundancy that characterize vertebrate gene regulation and increasing the difficulty of the task to understand their role and model Hox gene regulation. The current transgenic tools are well adapted to define the "activating" role of individual elements, but we would need to develop novel strategies, such as BAC transgenesis and other large scale approaches, to have access to the "repressive" roles of these elements and also understand communality effects, such as synergy, redundancy and complementarity.

Because of their striking conservation and role in body patterning, the evolution of Hox genes—or the role of Hox genes in evolution—has always been a highly intriguing question. It may well be very difficult to identify the ancestral regulatory mechanisms that controlled the urbilaterian Hox genes, but the sequencing of the genomes of additional chordates and animals from other phyla will undoubtedly help to clarify pending issues such as the origin of global regulatory elements, the evolution of gene numbers and regulatory elements within the clusters. However, as it is still hard to predict functional (dis)similarities from sequence comparison, this evo-devo approach will only bring new lights if it is complemented by functional analyses. This will require both the development of transgenic approaches in novel animal models, as well as the refinement and improvement of the transgenic assays themselves, to capture more accurately small expression changes that may correspond to important phenotypic transformations.

References

1. Lewis EB. A gene complex controlling segmentation in Drosophila. Nature 1978; 276:565-70.
2. McGinnis W, Levine MS, Hafen E et al. A conserved DNA sequence in homoeotic genes of the Drosophila Antennapedia and bithorax complexes. Nature 1984; 308:428-33.
3. Scott MP, Weiner AJ. Structural relationships among genes that control development: sequence homology between the Antennapedia, Ultrabithorax and fushi tarazu loci of Drosophila. Proc Natl Acad Sci USA 1984; 81:4115-9.
4. Duboule D, Baron A, Mahl P et al. A new homeo-box is present in overlapping cosmid clones which define the mouse Hox-1 locus. Embo J 1986; 5:1973-80.
5. Boncinelli E, Somma R, Acampora D et al. Organization of human homeobox genes. Hum Reprod 1988; 3:880-6.
6. Akam M. Hox and HOM: homologous gene clusters in insects and vertebrates. Cell 1989; 57:347-9.
7. Gaunt SJ, Krumlauf R, Duboule D. Mouse homeo-genes within a subfamily, Hox-1.4, -2.6 and -5.1, display similar anteroposterior domains of expression in the embryo, but show stage- and tissue-dependent differences in their regulation. Development 1989; 107:131-41.
8. Graham A, Papalopulu N, Krumlauf R. The murine and Drosophila homeobox gene complexes have common features of organization and expression. Cell 1989; 57:367-78.
9. Krumlauf R. Hox genes in vertebrate development. Cell 1994; 78:191-201.
10. Deschamps J. Ancestral and recently recruited global control of the Hox genes in development. Curr Opin Genet Dev 2007; 422-7.
11. Duboule D. Vertebrate Hox gene regulation: clustering and/or colinearity? Curr Opin Genet Dev 1998; 8:514-8.
12. Ikuta T, Yoshida N, Satoh N et al. Ciona intestinalis Hox gene cluster: Its dispersed structure and residual colinear expression in development. Proc Natl Acad Sci USA 2004; 101:15118-23.
13. Lemons D, McGinnis W. Genomic evolution of Hox gene clusters. Science 2006; 313:1918-22.
14. Seo HC, Edvardsen RB, Maeland AD et al. Hox cluster disintegration with persistent anteroposterior order of expression in Oikopleura dioica. Nature 2004; 431:67-71.
15. Renucci A, Zappavigna V, Zakany J et al. Comparison of mouse and human HOX-4 complexes defines conserved sequences involved in the regulation of Hox-4.4. Embo J 1992; 11:1459-68.
16. Tuggle CK, Zakany J, Cianetti L et al. Region-specific enhancers near two mammalian homeo box genes define adjacent rostrocaudal domains in the central nervous system. Genes Dev 1990; 4:180-9.
17. Whiting J, Marshall H, Cook M et al. Multiple spatially specific enhancers are required to reconstruct the pattern of Hox-2.6 gene expression. Genes Dev 1991; 5:2048-59.
18. Zakany J, Tuggle CK, Patel MD et al. Spatial regulation of homeobox gene fusions in the embryonic central nervous system of transgenic mice. Neuron 1988; 1:679-91.
19. Puschel AW, Balling R, Gruss P. Separate elements cause lineage restriction and specify boundaries of Hox-1.1 expression. Development 1991; 112:279-87.
20. Wada H, Garcia-Fernandez J, Holland PW. Colinear and segmental expression of amphioxus Hox genes. Dev Biol 1999; 213:131-41.
21. Shippy TD, Ronshaugen M, Cande J et al. Analysis of the Tribolium homeotic complex: insights into mechanisms constraining insect Hox clusters. Dev Genes Evol 2008; 127-39.
22. Irvine SQ, Martindale MQ. Expression patterns of anterior Hox genes in the polychaete Chaetopterus: correlation with morphological boundaries. Dev Biol 2000; 217:333-51.
23. Duboule D. The rise and fall of Hox gene clusters. Development 2007; 2549-60.
24. Kmita M, Duboule D. Organizing axes in time and space; 25 years of colinear tinkering. Science 2003;331-3.

25. Simeone A, Acampora D, Arcioni L et al. Sequential activation of HOX2 homeobox genes by retinoic acid in human embryonal carcinoma cells. Nature 1990; 346:763-6.
26. Papalopulu N, Lovell-Badge R, Krumlauf R. The expression of murine Hox-2 genes is dependent on the differentiation pathway and displays a collinear sensitivity to retinoic acid in F9 cells and Xenopus embryos. Nucleic Acids Res 1991; 19:5497-506.
27. Marshall H, Nonchev S, Sham MH et al. Retinoic acid alters hindbrain Hox code and induces transformation of rhombomeres 2/3 into a 4/5 identity. Nature 1992; 360:737-41.
28. Langston AW, Gudas LJ. Retinoic acid and homeobox gene regulation. Curr Opin Genet Dev 1994; 4:550-5.
29. Marshall H, Studer M, Popperl H et al. A conserved retinoic acid response element required for early expression of the homeobox gene Hoxb-1. Nature 1994; 370:567-71.
30. Zhang F, Popperl H, Morrison A et al. Elements both 5′ and 3′ to the murine Hoxd4 gene establish anterior borders of expression in mesoderm and neurectoderm. Mech Dev 1997; 67:49-58.
31. Oosterveen T, Niederreither K, Dollé P et al. Retinoids regulate the anterior expression boundaries of 5′ Hoxb genes in posterior hindbrain. EMBO J 2003; 262-9.
32. Oosterveen T, van Vliet P, Deschamps J et al. The direct context of a Hox retinoic acid response element is crucial for its activity. J Biol Chem 2003; 278:24103-7.
33. Gould A, Itasaki N, Krumlauf R. Initiation of rhombomeric Hoxb4 expression requires induction by somites and a retinoid pathway. Neuron 1998; 21:39-51.
34. Roelen BA, de Graaff W, Forlani S et al. Hox cluster polarity in early transcriptional availability: a high order regulatory level of clustered Hox genes in the mouse. Mech Dev 2002; 81-90.
35. Kmita M, van der Hoeven F, Zákány J et al. Mechanisms of Hox gene colinearity: transposition of the anterior Hoxb1 gene into the posterior HoxD complex. Genes Dev 2000; 198-211.
36. Manzanares M, Wada H, Itasaki N et al. Conservation and elaboration of Hox gene regulation during evolution of the vertebrate head. Nature 2000; 408:854-7.
37. Duboule D. Temporal colinearity and the phylotypic progression: a basis for the stability of a vertebrate Bauplan and the evolution of morphologies through heterochrony. Dev Suppl 1994; 135-42.
38. van der Hoeven F, Zákány J, Duboule D. Gene transpositions in the HoxD complex reveal a hierarchy of regulatory controls. Cell 1996; 1025-35.
39. Kondo T, Duboule D. Breaking colinearity in the mouse HoxD complex. Cell 1999; 407-17.
40. Forlani S, Lawson KA, Deschamps J. Acquisition of Hox codes during gastrulation and axial elongation in the mouse embryo. Development 2003; 3807-19.
41. Chambeyron S, Bickmore WA. Chromatin decondensation and nuclear reorganization of the HoxB locus upon induction of transcription. Genes Dev 2004; 18:1119-30.
42. Chambeyron S, Da Silva NR, Lawson KA et al. Nuclear re-organisation of the HoxB complex during mouse embryonic development. Development 2005; 132:2215-23.
43. Morey C, Da Silva NR, Perry P et al. Nuclear reorganisation and chromatin decondensation are conserved, but distinct, mechanisms linked to Hox gene activation. Development 2007; 909-19.
44. Morey C, Da Silva NR, Kmita M et al. Ectopic nuclear reorganisation driven by a Hoxb1 transgene transposed into HoxD. J Cell Sci 2008; 571-7.
45. Sharpe J, Nonchev S, Gould A et al. Selectivity, sharing and competitive interactions in the regulation of Hoxb genes. Embo J 1998; 17:1788-98.
46. Gould A, Morrison A, Sproat G et al. Positive cross-regulation and enhancer sharing: two mechanisms for specifying overlapping Hox expression patterns. Genes Dev 1997; 11:900-13.
47. Gerard M, Chen JY, Gronemeyer H et al. In vivo targeted mutagenesis of a regulatory element required for positioning the Hoxd-11 and Hoxd-10 expression boundaries. Genes Dev 1996; 10:2326-34.
48. Mann RS. Why are Hox genes clustered? Bioessays 1997; 19:661-4.
49. Zakany J, Duboule D. Hox genes in digit development and evolution. Cell Tissue Res 1999; 296:19-25.
50. Zakany J, Duboule D. Hox genes and the making of sphincters. Nature 1999; 401:761-2.
51. Dolle P, Izpisua-Belmonte JC, Falkenstein H et al. Coordinate expression of the murine Hox-5 complex homoeobox-containing genes during limb pattern formation. Nature 1989; 342:767-72.
52. Dolle P, Izpisua-Belmonte JC, Boncinelli E et al. The Hox-4.8 gene is localized at the 5′ extremity of the Hox-4 complex and is expressed in the most posterior parts of the body during development. Mech Dev 1991; 36:3-13.
53. Nelson CE, Morgan BA, Burke AC et al. Analysis of Hox gene expression in the chick limb bud. Development 1996; 122:1449-66.
54. Spitz F, Gonzalez F, Peichel C et al. Large scale transgenic and cluster deletion analysis of the HoxD complex separate an ancestral regulatory module from evolutionary innovations. Genes Dev 2001;2209-14.

55. Lee AP, Koh EG, Tay A et al. Highly conserved syntenic blocks at the vertebrate Hox loci and conserved regulatory elements within and outside Hox gene clusters. Proc Natl Acad Sci USA 2006; 6994-9.
56. Spitz F, Montavon T, Monso-Hinard C et al. A t(2; 8) balanced translocation with breakpoints near the human HOXD complex causes mesomelic dysplasia and vertebral defects. Genomics, 2002; 493-8.
57. Dlugaszewska B, Silahtaroglu A, Menzel C et al. Breakpoints around the HOXD cluster result in various limb malformations. J Med Genet 2006; 111-8.
58. Spitz F, Herkenne C, Morris MA et al. Inversion-induced disruption of the Hoxd cluster leads to the partition of regulatory landscapes. Nat Genet 2005; 889-93.
59. Di-Poï N, Zákány J, Duboule D. Distinct roles and regulations for HoxD genes in metanephric kidney development. PLoS Genet 2007; e232.
60. Gonzalez F, Duboule D, Spitz F. Transgenic analysis of Hoxd gene regulation during digit development. Dev Biol 2007; 306(2):847-59.
61. Spitz F, Gonzalez F, Duboule D. A global control region defines a chromosomal regulatory landscape containing the HoxD cluster. Cell 2003; 405-17.
62. Carvajal JJ, Cox D, Summerbell D et al. A BAC transgenic analysis of the Mrf4/Myf5 locus reveals interdigitated elements that control activation and maintenance of gene expression during muscle development. Development 2001; 128:1857-68.
63. Uchikawa M, Ishida Y, Takemoto T et al. Functional analysis of chicken Sox2 enhancers highlights an array of diverse regulatory elements that are conserved in mammals. Dev Cell 2003; 4:509-19.
64. Monge I, Kondo T, Duboule D. An enhancer-titration effect induces digit-specific regulatory alleles of the HoxD cluster. Dev Biol 2003; 256:212-20.
65. Kmita M, Tarchini B, Duboule D et al. Evolutionary conserved sequences are required for the insulation of the vertebrate HoxD complex in neural cells. Development 2002; 5521-8.
66. Kmita M, Fraudeau N, Hérault Y et al. Serial deletions and duplications suggest a mechanism for the collinearity of Hoxd genes in limbs. Nature 2002; 145-50.
67. Favier B, Le Meur M, Chambon P et al. Axial skeleton homeosis and forelimb malformations in Hoxd-11 mutant mice. Proc Natl Acad Sci USA 1995; 92:310-4.
68. Davis AP, Capecchi MR. Axial homeosis and appendicular skeleton defects in mice with a targeted disruption of Hoxd-11. Development 1994; 120:2187-98.
69. Kondo T, Dollé P, Zákány J et al. Function of posterior Hoxd genes in the morphogenesis of the anal sphincter. Development 1996; 2651-9.
70. Duboule D, Morata G. Colinearity and functional hierarchy among genes of the homeotic complexes. Trends Genet 1994; 10:358-64.
71. Palstra RJ, de Laat W, Grosveld F. Chapter 4 beta-globin regulation and long-range interactions. Adv Genet 2008; 107-42.
72. Montavon T, Le Garrec JF, Kerszberg M et al. Modeling Hox gene regulation in digits: reverse collinearity and the molecular origin of thumbness. Genes Dev 2008; 346-59.
73. Tarchini B, Duboule D. Control of Hoxd genes' collinearity during early limb development. Dev Cell 2006; 93-103.
74. Bergeron J, Clappier E, Cauwelier B et al. HOXA cluster deregulation in T-ALL associated with both a TCRD-HOXA and a CALM-AF10 chromosomal translocation. Leukemia 2006; 20:1184-7.
75. Poppe B, Yigit N, De Moerloose B et al. HOXA gene cluster rearrangement in a t(7; 9)(p15; q34) in a child with MDS. Cancer Genet Cytogenet 2005; 162:82-4.
76. Yue Y, Farcas R, Thiel G et al. De novo t(12; 17)(p13.3; q21.3) translocation with a breakpoint near the 5′ end of the HOXB gene cluster in a patient with developmental delay and skeletal malformations. Eur J Hum Genet 2007; 15:570-7.
77. McEwen GK, Woolfe A, Goode D et al. Ancient duplicated conserved noncoding elements in vertebrates: a genomic and functional analysis. Genome Res 2006; 16:451-65.
78. Lehoczky JA, Williams ME, Innis JW. Conserved expression domains for genes upstream and within the HoxA and HoxD clusters suggests a long-range enhancer existed before cluster duplication. Evol Dev 2004; 423-30.
79. Suster ML, Kania A, Liao M et al. A novel conserved evx1 enhancer links spinal interneuron morphology and cis-regulation from fish to mammals. Dev Biol 2009; 325:422-33.
80. Sordino P, van der Hoeven F, Duboule D. Hox gene expression in teleost fins and the origin of vertebrate digits. Nature 1995; 375:678-81.
81. Davis MC, Dahn RD, Shubin NH. An autopodial-like pattern of Hox expression in the fins of a basal actinopterygian fish. Nature 2007; 473-6.
82. Freitas R, Zhang G, Cohn MJ. Biphasic Hoxd gene expression in shark paired fins reveals an ancient origin of the distal limb domain. PLoS ONE 2007; e754.
83. Ahn D, Ho RK. Tri-phasic expression of posterior Hox genes during development of pectoral fins in zebrafish: implications for the evolution of vertebrate paired appendages. Dev Biol 2008; 322:220-33.

84. Amemiya CT, Prohaska SJ, Hill-Force A et al. The amphioxus Hox cluster: characterization, comparative genomics and evolution. J Exp Zoolog B Mol Dev Evol 2008; 310:465-77.
85. Kmita M, Kondo T, Duboule D. Targeted inversion of a polar silencer within the HoxD complex re-allocates domains of enhancer sharing. Nat Genet 2000; 26:451-4.
86. McIntyre DC, Rakshit S, Yallowitz AR et al. Hox patterning of the vertebrate rib cage. Development 2007; 134:2981-9.
87. Wellik DM, Capecchi MR. Hox10 and Hox11 genes are required to globally pattern the mammalian skeleton. Science 2003; 301:363-7.
88. Maconochie MK, Nonchev S, Studer M et al. Cross-regulation in the mouse HoxB complex: the expression of Hoxb2 in rhombomere 4 is regulated by Hoxb1. Genes Dev 1997; 11:1885-95.
89. Tumpel S, Cambronero F, Ferretti E et al. Expression of Hoxa2 in rhombomere 4 is regulated by a conserved cross-regulatory mechanism dependent upon Hoxb1. Dev Biol 2007; 302:646-60.
90. Yekta S, Tabin CJ, Bartel DP. MicroRNAs in the Hox network: an apparent link to posterior prevalence. Nat Rev Genet 2008; 9:789-96.
91. Rinn JL, Kertesz M, Wang JK et al. Functional demarcation of active and silent chromatin domains in human HOX loci by noncoding RNAs. Cell 2007; 129:1311-23.
92. Visel A, Blow MJ, Li Z et al. ChIP-seq accurately predicts tissue-specific activity of enhancers. Nature 2009; 457:854-8.
93. Yamagishi T, Ozawa M, Ohtsuka C et al. Evx2-Hoxd13 intergenic region restricts enhancer association to Hoxd13 promoter. PLoS ONE 2007; 2:e175.

SECTION II

Evolution of Hox Genes and Complexes

Chapter 5

The Early Evolution of Hox Genes: A Battle of Belief?

Bernd Schierwater* and Kai Kamm

Abstract

For more than a century the origin of metazoan animals and for less than three years the early evolution of Hox genes has been debated. Both discussions are intrinsically tied together. New data from whole genome sequencing and recent progress in phylogeny of basal metazoans allow to provide an answer. The evolution of diploblastic animals (Placozoa, Porifera, Ctenophora and Cnidaria) and Bilateria (all higher animals) went parallel. The early split of these two lineages led to the evolution of a Hox system in Bilateria and the presence of Hox-like genes in Cnidaria and Placozoa.

The Hox System

Until three years ago the presence of a so-called "Hox system" was believed to present a genetic synapomorphy uniting all metazoans and separating them from Protozoa (c.f. refs. 1-3). The Hox system was defined upon the presence, organization and expression of a cluster of certain homeotic genes that define segment or region identity along an anterior-posterior (A-P) axis (for review see ref. 4). The Hox cluster has been the Rosetta Stone of comparative developmental biology and added tremendously to our understanding of bauplan development in higher animals. In bilaterians Hox genes are characteristically organized in clusters whose genomic organization directly reflects spatial and temporal expression along the A-P axis. This pattern of organization is functionally important and has led to the assumption that much of the morphological variation seen across the animal kingdom can be directly attributed to different numbers of Hox genes or differential usage of the Hox system.[5-7] In 2006, Kamm et al[8] defined a "canonical Hox system" as a set of closely linked and interacting homeobox genes that are directly related to the Hox classes of *Drosophila* and mammals and that, through their combined actions, are primarily responsible for patterning most or all tissues along the anterior-posterior body axis (cf. refs. 3,4,9). This definition became necessary when comparing data from Bilateria to those from diploblastic animals, particularly Cnidaria. Hox-like genes have been identified in a wide variety of cnidarians (e.g., ref. 10) but, in contrast to a number of other key regulatory gene types, their status is often equivocal. When Kamm et al (2006)[8] examined the structure, genomic organization and expression of these genes in both a hydrozoan and anthozoan cnidarian, the presence of a Hox system in Cnidaria became all of a sudden questionable. By weakening the definition of a Hox system and speculating about homologies two subsequent studies by Chourrout et al (2006)[11] and Ryan et al (2007)[12] tried to rescue the idea of a general Hox system in Metazoa. The latter attempts seem to be futile given most recent information from whole genome data and progress in phylogenetic analyses at the base of metazoan evolution.

*Corresponding Author: Bernd Schierwater—ITZ, Ecology and Evolution, Tierärztliche Hochschule Hannover, Bünteweg 17d, D-30559 Hannover, Germany. Email: bernd.schierwater@ecolevol.de

Hox Genes: Studies from the 20th to the 21st Century, edited by Jean S. Deutsch.

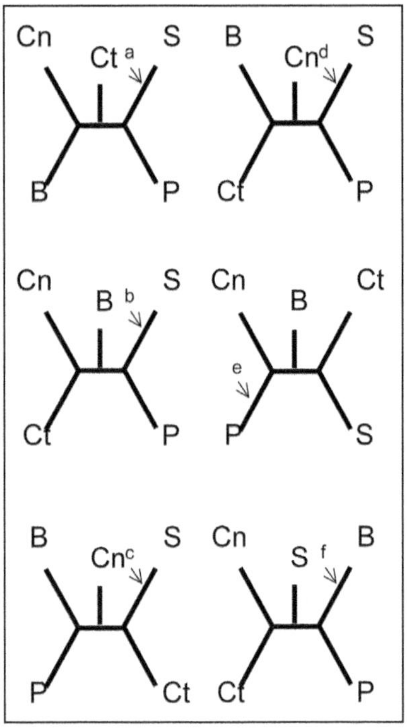

Figure 1. Discussed Relationships at the Base of the Metazoan Tree. Potential arrangements of five critical taxa (B, Bilateria; Cn, Cnidaria; Ct, Ctenophora; P, Placozoa; and S, Porifera) are shown. Arrows indicate the root of the networks. The lowercase letters refer to publications in Table S1 in Schierwater et al 2009[13] that support the root for trees with all five taxa. From Schierwater et al 2009.[13]

Phylogenetic Evidence

In order to unravel the early evolution of Hox and Hox-like genes, knowledge of phylogenetic relationships near the base of the metazoan tree of life becomes a prerequisite. These relationships have been highly controversial and a large number of conflicting phylogenetic scenarios have been suggested (cf. refs. 13,14).

The most comprehensive study yet available analyzed the sum of morphological evidence, the secondary structure of mitochondrial ribosomal genes and molecular sequence data from mitochondrial and nuclear genes that amass over 9000 phylogenetically informative characters from 24 to 73 taxa.[13] Together with mtDNA genome data[15] and Hox-like gene expression patterns, these data provide strong evidence that Placozoa (*Trichoplax adhaerens*)[16] are basal relative to all other diploblast phyla and that diploblastic animals and Bilateria are sister groups.[17] This unusual hypothesis is surprising, yet it is not new. Several recent studies, although based on smaller data sets and smaller taxon samplings, have suggested an early split between diploblastic animals and Bilateria.[13,17] Basically all plausible possibilities for the arrangement of Bilateria relative to the four diploblast phyla (Cnidaria, Ctenophora, Porifera and Placozoa) have been suggested by recent molecular analyses. Figure 1 shows six plausible scenarios for the relationships of five taxonomic groups (Bilateria, Cnidaria, Ctenophora, Porifera and Placozoa). For five taxa and an outgroup there are 105 ways to arrange these taxa in dichotomous branching trees. 99 of these trees can be eliminated as not plausible leaving us with six possible hypotheses. Remarkably, all six have been suggested in publications in the last two years alone. For instance, Srivastava et al (2008)[18] suggest Placozoa as the sister group to both Cnidaria and Bilateria with sponges branching off earlier.

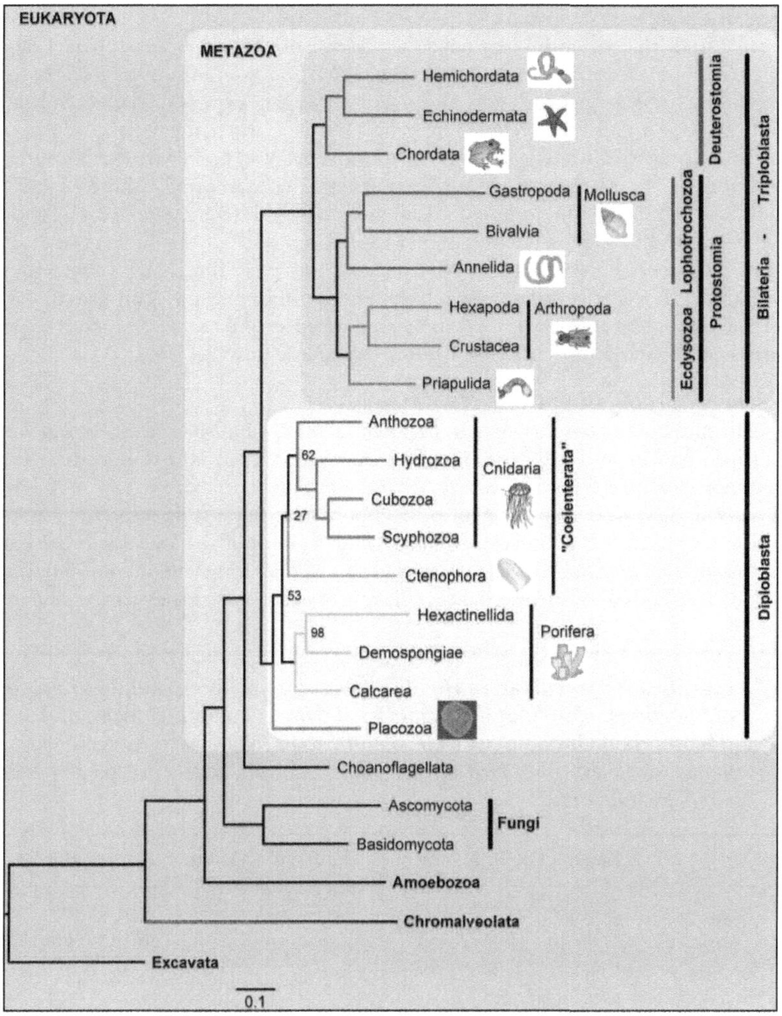

Figure 2. Current knowledge suggests that diploblastic animals (including Coelenterata, Placozoa and Porifera) are sister groups to Bilateria. According to this hypothesis, the evolution of Hox genes in Bilateria and Hox-like genes in diploblasts went parallel. From Schierwater et al 2009.[13]

Another recent study suggests a basal position for Ctenophora and Anthozoa and that Cnidaria are not sister to Bilateria, but rather to Porifera.[19] Unfortunately this study is useless in the given context, since it does not include Placozoa in the analysis. Ruiz-Trillo et al (2008),[20] a study that includes Placozoa, also suggests that Bilateria and Placozoa are sister groups. Several recent analyses of mitochondrial genome sequence data[15,21,22] place Bilateria as sister to all nonBilateria with Placozoa as the most basal diploblast (Fig. 2). For overview, details and complete references see references 13, 17 and 23.

Altogether, trying to look into the mess of different hypotheses, the sum of current knowledge suggests that Bilateria are the sister group to either Cnidaria, or to all other diploblasts and that Placozoa are basal within the diploblasts. This implies that Hox genes in Bilateria evolved parallel to Hox-like genes in Cnidaria and other diploblastic animals.

Opposing Views

The first study that seriously questioned the presence of a Hox system in diploblastic animals was based on empirical data from both, an anthozoan (*Nematostella*) and a hydrozoan (*Eleutheria*).[8] The authors report gene structure, genomic organization, gene expression data and phylogenetic analyses for a large number of potential Hox genes. The data refused the criteria of cluster organization, colinear expression and gene homology for the majority of cnidarian genes and thus the preconditions for the existence of a Hox system in Cnidaria. Interestingly, shortly thereafter two other studies used mainly the anthozoan data, used slightly different criteria for assigning gene homology and slightly different definition of a Hox system and concluded that there was a Hox system in Cnidaria (Chourrout et al 2006, Ryan et al 2007[11,12]). There can be numerous reasons why different authors derive different conclusions from similar or even identical data sets (cf. ref. 24). In the following we aim to list and compare differences in data analyses and interpretation between the three above papers as objective and traceable as possible.

Differences in Assigning Gene Homology

Several crucial homology assignments of potential Hox or Hox-like genes, respectively, differ among papers (see Tables 1 and 2 for a list of relevant genes). There is no dispute that Cnidarians have anterior Hox-like (*Cnox-5*, *anthox6/6a*) and *Gsx* genes and that the new *anthox9* gene is highly derived. But three major differences are important to note. With respect to putative posterior Hox/*Cdx*-like genes Kamm et al[8] found candidates in the hydrozoan *Eleutheria* but not in the anthozoan *Nematostella*. The Chourrout et al paper[11] does not mention some published data from hydrozoans and comes to the conclusion that there were no posterior genes in Cnidaria. Ryan

Table 1. *Classification of cnidarian Hox-like genes into different families in recent publications. Note that the Kamm et al dataset contained hydrozoan (*Eleutheria dichotoma*) as well as anthozoan (*Nematostella vectensis*) genes while the other two studies analyzed anthozoan Hox-like genes from Nematostella only.*

	Kamm et al 06	Chourrout et al 06	Ryan et al 07
Anterior type	*Anthox6/6a* (Nv), *Cnox-5* (Ed)	*Anthox6/6a* (Nv)	*Anthox6/6a* (Nv), *Anthox7/8/8a* (Nv)
Posterior type	Cnox-3, -4 (Ed)—referred to as posterior Hox/*Cdx*-like genes	None	*Anthox1/1a* (Nv)
Hox3 type	None	*Anthox7/8/8a*	None
Gsx type	Anthox-2 (Nv), Cnox-2 (Ed)	Anthox-2 (Nv)	Anthox-2 (Nv)
Cdx type	Cnox-4 (Ed)—referred to as posterior Hox/*Cdx*-like gene	*NvHD065*—referred to as chimera between *Xlox/Cdx*	*NvHD065*—referred to as chimera between *Xlox/Cdx*
Xlox type	None	*NvHD065*—referred to as chimera between *Xlox/Cdx*	*NvHD065*—referred to as chimera between *Xlox/Cdx*
Genuine cnidarian Hox-like genes	-The paralogs *anthox7/8/8a* (Nv) -*Anthox1/1a* (Nv) -*Anthox9* (Nv) -*Cnox-1* (Ed)	-*Anthox1/1a* (Nv) -*Anthox9* (Nv)	-*Anthox9* (Nv)

et al[12] is the only paper suggesting anthox1/1a as a possible posterior gene. According to Kamm et al[8] other cnidarian Hox-like genes are not assignable to true Hox classes and thus most likely are genuine cnidarian genes. In contrast the paralogous genes *anthox7/8/8a* are believed to be similar to the *Hox3* group[11] or to the *Hox2* group.[12] Only the latter two studies report a NvHD065 homeodomain, which is regarded as an ancient *Xlox* in Chourrout et al[11] or as the predecessor of both *Cdx* and *Xlox* (Ryan et al 2007[12]). The critical reader should take a look at the trees that are supposed to support these interpretations. In a newer study the *NvHD0065* gene shows relationships to *Cdx*.[25] The above differences in assigning gene homology to potential Hox or Hox-like genes lead to astonishing different conclusions in all three papers.

The Kamm et al study[8] concludes that the cnidarian bilaterian ancestor (CBA) possessed the ancestors of anterior Hox and posterior Hox/*Cdx* genes and a *Gsx* gene. The remaining cnidarian genes most likely postdate the cnidarian/bilaterian split. In short, there is no Hox system in Cnidaria. The Chourrout et al study[11] concludes that the cnidarian/bilaterian ancestor possessed anterior and group 3 Hox genes and the corresponding ParaHox genes (*Gsx* and *Xlox*), which would be congruent with a minimal Hox system. The Ryan et al[12] paper pushes towards the presence of a Hox system in Cnidaria and suggests that the cnidarian bilaterian ancestor possessed the ancestors of anterior Hox and posterior Hox genes, *Gsx* and a chimera between *Xlox* and *Cdx* from which both originated.

The Kamm et al study missed one *Nematostella* Hox-like gene, NvHD065. If this gene had been included it would not have changed the picture, however. The other two studies leave out all comparative data from Hydrozoa and make some homology assignments that need to be judged carefully. If one accepts the suggested homologies, it may still not be obvious to many readers how this would support a true Hox system. It is unquestioned that all phylogenetic analyses failed to identify clear orthologs to all four bilaterian Hox classes (anterior, group 3, central, posterior). The assignment of *anthox7/8/8a* to *Hox3* and the new HD to *Xlox* is far from certain and also not adopted from Ryan et al.[12] Moreover, Ryan et al agree that the new HD is close to *Cdx*. If this was true the conclusion should be that cnidarians have posterior Hox/*Cdx*-like genes, as suggested by Kamm et al.[8]

How likely is it that the suggested *Xlox*/*Cdx* chimera (Ryan et al 2007[12]) is the ancestor of both, *Xlox* and *Cdx*? If we imagine a *cis*-duplication of this gene, one copy must have lost *Xlox* characters and one must have lost *Cdx* characters. A more parsimonious interpretation would be

Table 2. Nomenclature of Nematostella Hox-like genes in recent publications (see text for homologies and explanations)

Kamm et al 06	Chourrout et al 06	Ryan et al 07
Anthox1	HoxF	*Anthox1*
Anthox1a	HoxE	*Anthox1a*
Anthox2	*GSX*	Anthox2
Anthox6	HoxA	*Anthox6*
Anthox6a	HOXB	*Anthox6a*
Anthox7	HoxC	*Anthox7*
Anthox8	HoxDa	*Anthox8a*
Anthox8a	HoxDb	*Anthox8b*
Anthox9	HOXR	*Anthox9*
Antheve	EVX	Evx
	XLOX/CDX	NVHD065

that this gene represents a derived posterior Hox/*Cdx*-like gene. The latter would be consistent with a recent and more complete ANTP gene analyses.[25] Besides the above differences in analyses and interpretation, unfortunately, further confusion arises from wrong or misleading quotations. For example, one paper says that if there is no true Hox system one cannot have true Hox genes in the original meaning.[8] Another paper takes those words out of their context and wrongly cites the former paper as rejecting a common origin for all cnidarian/bilaterian Hox/ParaHox genes.

Linkage Data

It might be thought that linkage is nothing to argue about rather than a simple observation on gene order along a chromosome. In praxis, however, there are often possibilities to complicate things. If one disagrees on gene homology one may also disagree on linkage as a result. If one disagrees on the definition of linkage, e.g., allowed physical distance and separation by other genes between two "linked" genes, one will obviously also disagree on linkage interpretation. With respect to linkage of Hox-like genes in Cnidaria the reader observes a few, however quite crucial differences between different papers.

When comparing the data sets the reader will notice that the two papers that report data from *Nematostella* only, provide the most linkage data by also including putative ParaHox genes. The third paper uses a smaller *Nematostella* data set but provides a second data set from a hydrozoan. The reported linkage data are shown in Figure 3. Kamm et al[8] conclude that—with the exception of independently duplicated genes (*anthox7/8/8a*)—the cnidarian genes are not linked rather than flanked over great distance by unrelated genes. This observation contradicts an ancient Hox cluster. The linkage of *anthox1a* and *anthox9* is no sign of an ancient Hox cluster, since these genes are no orthologs of true Hox genes (*anthox9* might even be a pseudogene). Quite differently, Chourrout et al regard the *anthox7/8/8a* as related by homology to *Hox3*.[11] With this interpretation they found additional linkage of *anthox7/8/8a* to antheve and *anthox6* and regard this as an early cluster of anterior and group3 Hox genes. Interestingly, Ryan et al[12] see *anthox7/8/8a* as anterior *Hox2* homolog. In this interpretation we had a cluster of four anterior Hox-like genes, which is congruent with the interpretation in Kamm et al.[8] The problem with the "early cluster of anterior and group3 Hox genes scenario" is that more comprehensive molecular analyses did not identify *anthox7/8/8a* as related to *Hox3* and a relationship to other anterior Hox genes is not well supported.[8,12,25] Nonetheless two of the three papers conclude that Cnidaria have anterior and posterior gene types.

If the CBA had anterior and posterior genes these must have been linked, if they are the ancestors of true Hox genes. This linkage could be interpreted as fulfilling one of the three criteria for

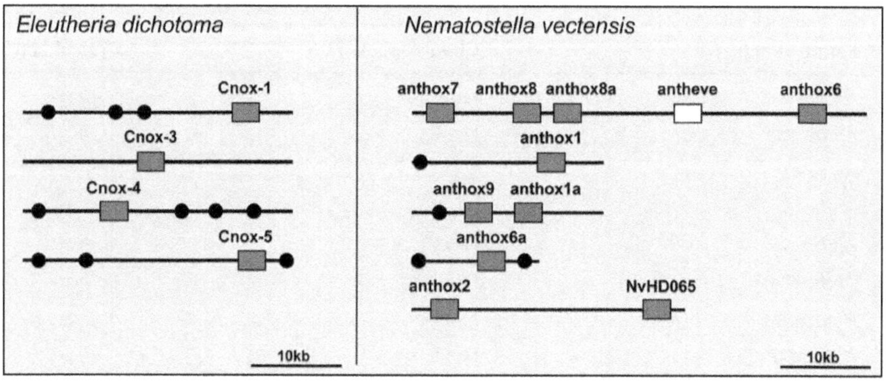

Figure 3. Genomic organization of cnidarian Hox-like genes in the hydrozoan *Eleutheria dichotoma* and the anthozoan *Nematostella vectensis*. Black circles denote unrelated genes. For details see references 8, 11 and 12.

a Hox system. Unfortunately neither the cnidarian Hox-like genes nor their function are directly comparable to genes of a true Hox cluster. Thus two of the three criteria for a true (canonical) Hox system are not fulfilled.

Since some people seem to get confused with ANTP vs. Hox genes, one should note, that the linkage of *anthox6* and *even-skipped* and the further linkage of other ANTP genes to them (e.g., Mnx and Rough) is the remnant of an ancient ANTP array[8,12,25] and probably has nothing to do with an ancient Hox system, it rather predates it.

Expression Data

One of the papers suggesting a Hox system in Cnidaria[11] had to draw conclusions in the absence of any information on expression data, i.e., without the third criterion for a Hox system. The other paper supporting the Hox view (Ryan et al[12]) incorporates expression data from the anthozoan *Nematostella* (see Fig. 4A,B), while the contrary view uses both expression data from *Nematostella* as well as from the hydrozoan *Eleutheria dichotoma.*[8]

Most important seems to be the discussion of *Cnox-5* and its putative homologs in Anthozoa and Hydrozoa. *Anthox6* (anterior) is expressed oral in the polyp (= posterior in the planula with respect to swimming direction) in *Nematostella,* while its homologs are expressed differently in hydrozoans (e.g., *Cnox-5ed*[8] and *Podocoryne Cnox-1pc*;[26] Fig. 4C). *Anthox6a* (also anterior Hox-like) is expressed along the body column.[12]

The Hox like genes *EdCnox-1* (ortholog to NvAnthox1/1a) and *EdCnox-3* (no *Nematostella* ortholog) are also of interest here. Both are expressed in the hydrozoan medusa,[8] which is not directly comparable to the *Nematostella* data, since Anthozoa lack a medusa stage.

The observation that *Anthox 7/8/8a,* NvHD065 and *anthox1a* are expressed along the body column and *Anthox1* aborally in the polyp led some authors to conclude that *Nematostella* Hox genes pattern the primary body axis.[12] This seems very optimistic for the above reasons of uncertain homology and quite inconsistent expression patterns in Cnidaria. In addition one might notice that in a radially symmetric organism with an oral-aboral body axis any gene expression will be somewhere along the axis so that *cum grano salis* any gene is expressed in relation to the axis. This way it can hardly be avoided that expression of putative anterior (*anthox7/8/8a* and *anthox6a*), putative posterior (*anthox1a*) and the *Xlox/Cdx* (NvHD065) chimera along the body column is related to the body axis. When compared to expression patterns in Bilateria one finds that the expression along the body column is not in overlapping patterns like in Bilateria. All authors agree that one anterior gene (*anthox6*) is expressed orally in *Nematostella.* The problem is that the orthologs in other Cnidaria are expressed differently. Moreover, even in one and the same species (*Nematostella*), the expression of closely related genes differs completely (i.e., genes of the same type like *anthox6* and 6a, anthox1 and 1a; see Fig. 4A,B)[12]—hence it does neither reflect conservation of function nor does this scheme follow the pattern of duplication and subfunctionalization (which is likely to have occurred during the expansion of the Hox cluster in Bilateria).

Conclusion

All of the above scenarios look more or less plausible. They differ in which and how many Hox or Hox-like genes are present in Cnidaria (see also ref. 27). A recent study even shows that a genomic region in a cnidarian (*Nematostella*) is syntenic to the ParaHox regions in Bilateria,[28] which makes it likely that the *cis-* or *trans-*duplication event that produced Hox and ParaHox regions has occurred before the split between cnidarians and bilaterians. Nevertheless, there is no dispute that cnidarians had the makings of a Hox system[8] and hence already possessed the appropriate genomic regions. The question should rather be if such a "region" was already developed to such an extent in Cnidaria that it was exploited for axial patterning comparable to Bilateria. That this seems not the case has been already suggested by Kamm et al[8] and has been confirmed recently by Chiori et al,[29] who compared new expression data of Hox-like genes in a hydrozoan to older data from the literature. They observed that: "Cross species comparison reveals a strong variability of gene expression along the oral-aboral axis and during the life cycle among cnidarian lineages. The

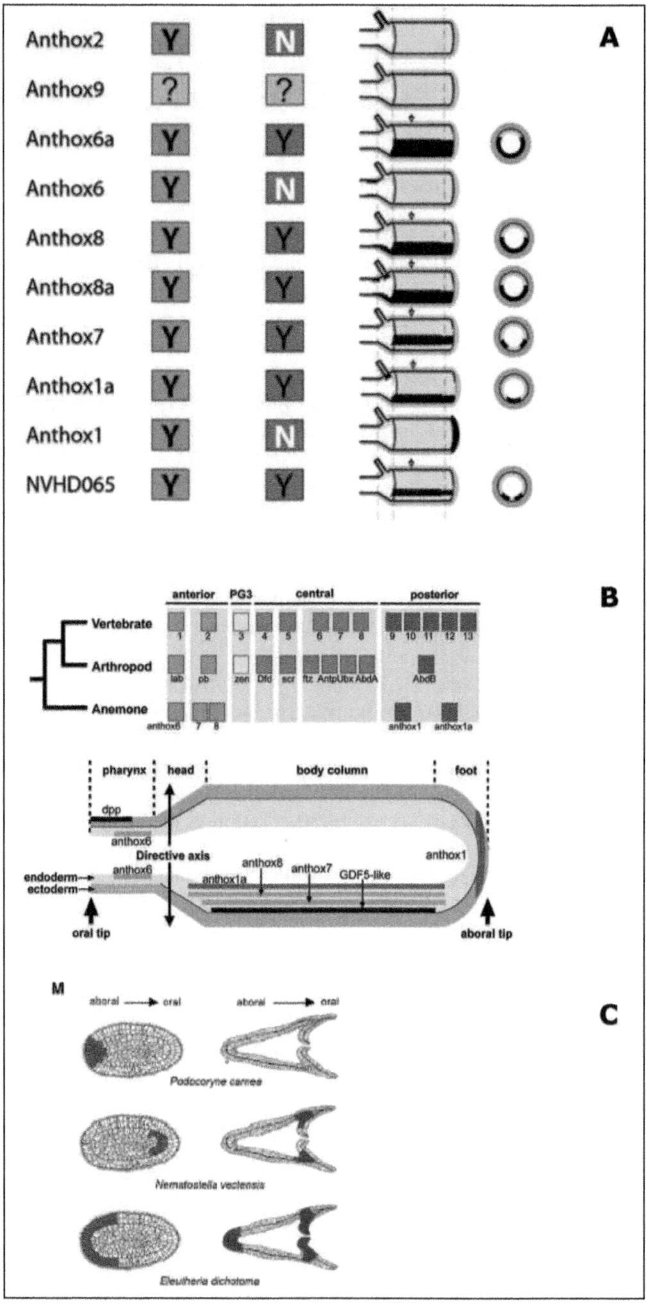

Figure 4. Expression patterns of cnidarian Hox-like genes: A,B) Expression patterns in the polyp of *Nematostella vectensis* according to Ryan et al 2007 (A)[12] and Finnerty et al 2004 (B).[37] B) Reprinted from reference 37 with permission from AAAS. C) Comparison of the expression of anterior Hox-like genes in the planula and the polyp of different cnidarians (*Cnox-1—Podocoryne; anthox6—Nematostella; Cnox-5—Eleutheria*).[8] Reprinted from reference 8 with permission from Elsevier. Further explanations in the text.

most parsimonious interpretation is that the Hox code, colinearity and conservative role along the antero-posterior axis are bilaterian innovations."[29] (In addition to Chiori et al see also Schierwater and Kuhn[30] and Cartright et al.[31])

Hence no scenario can obscure the obvious that Cnidaria do not possess a Hox system directly comparable to Bilateria—in terms of sequence identity to all four Hox classes, their (ancestral) linkage and a conserved role in patterning most or all tissues along the A-P (oral-aboral, O-A) axis during development. Even assuming extreme divergence, the situation in Cnidaria cannot easily be twisted to a derived homolog of a bilaterian Hox system—like is the case for example in urochordates.[32] Acoelomorph flatworms exhibit a very simple Hox system, which may represent the primitive bilaterian Hox condition,[33] but see reference 34 for discussion. It would be a logical gradual transition from a ProtoHox (*Gsx*) gene in Placozoa, loss of *Gsx* in Porifera, anterior and posterior gene types in Cnidaria, primitive Hox system in Acoelomorpha to an elaborate Hox system in Eubilateria.

All the controversial discussions should also be viewed with respect to the fact that in sharp contrast to Hox-like genes we have little problems to assign true orthologies to almost all nonHox ANTP superclass genes in Cnidaria.[35] This simple fact has important implications: It favors the view that the nonHox ANTP superclass gene families originated before the Hox genes and it contradicts the speculation that the difficulties to assign true orthologies to cnidarian Hox-like genes are the result of early divergence from the bilaterian lineage. The nonHox ANTP genes are presumably older and had more time to diverge and even in Placozoa and sponges we find true orthologs of nonHox ANTP genes.[25,36] The data from Placozoa and sponges show a better fit to the concept of independent evolution of Hox genes in Bilateria and Hox-like genes in diploblasts. In Placozoa we see a low diversity of ANTP genes and *Gsx* as the only Hox-like gene.[25] In Porifera we find only nonHox ANTP genes (loss of *Gsx*?)[36] and in Cnidaria an almost complete nonHox ANTP gene repertoire plus some orthologs to bilaterian Hox/ParaHox genes (*Gsx*, anterior, posterior Hox/*Cdx*-like).[8,11,12,35] Finally, there is still the unanswered question whether the O-A-axis is homologous to the A-P-axis, making any Hox system rescue attempts in Cnidaria even harder. Most recent phylogenetic data, which suggest an early split between Cnidaria and Bilateria, are congruent with a parallel evolution of Hox genes in the two lineages.[13]

References

1. Slack JM, Holland PW, Graham CF. The zootype and the phylotypic stage. Nature 1993; 361:490-2.
2. Miller DJ, Miles A. Homeobox genes and the zootype. Nature 1993; 365:215-6.
3. Schierwater B, Desalle R. Current problems with the zootype and the early evolution of Hox genes. J Exp Zool 2001; 291:169-74.
4. Garcia-Fernandez J. Hox, ParaHox, ProtoHox: facts and guesses. Heredity 2005; 94:145-52.
5. Gellon G, McGinnis W. Shaping animal body plans in development and evolution by modulation of Hox expression patterns. Bioessays 1998; 20:116-25.
6. Prince V. The Hox Paradox: More complex(es) than imagined. Dev Biol 2002; 249:1-15.
7. Amores A, Suzuki T, Yan YL, et al. Developmental roles of pufferfish Hox clusters and genome evolution in ray-fin fish. Genome Res 2004; 14:1-10.
8. Kamm K, Schierwater B, Jakob W et al. Axial patterning and diversification in the cnidaria predate the Hox system. Curr Biol 2006; 16:920-6.
9. Ferrier DE, Holland PW. Ancient origin of the Hox gene cluster. Nat Rev Genet 2001; 2:33-8.
10. Gauchat D, Mazet F, Berney C, et al. Evolution of Antp-class genes and differential expression of Hydra Hox/paraHox genes in anterior patterning. Proc Natl Acad Sci USA 2000; 97:4493-8.
11. Chourrout D, Delsuc F, Chourrout P, et al. Minimal ProtoHox cluster inferred from bilaterian and cnidarian Hox complements. Nature 2006; 442:684-7.
12. Ryan JF, Mazza ME, Pang K, et al. Pre-bilaterian origins of the Hox cluster and the Hox code: evidence from the sea anemone, Nematostella vectensis. PLoS ONE 2007; 2:e153.
13. Schierwater B, Eitel M, Jakob W, et al. Concatenated analysis sheds light on early metazoan evolution and fuels a modern Urmetazoon hypothesis. PLoS Biology 2009; 7:e20.
14. Blackstone NW. A new look at some old animals. PLoS Biol 2009; 7:e7.
15. Dellaporta SL, Xu A, Sagasser S, et al. Mitochondrial genome of Trichoplax adhaerens supports placozoa as the basal lower metazoan phylum. Proc Natl Acad Sci USA 2006; 103:8751-6.
16. Schierwater B. My favorite animal, Trichoplax adhaerens. Bioessays 2005; 27:1294-302.

17. Schierwater B, de Jong D, Desalle R. Placozoa and the evolution of Metazoa and intrasomatic cell differentiation. Int J Biochem Cell Biol 2009; 41:370-9.
18. Srivastava M, Begovic E, Chapman J, et al. The Trichoplax genome and the nature of placozoans. Nature 2008; 454:955-60.
19. Dunn CW, Hejnol A, Matus DQ, et al. Broad phylogenomic sampling improves resolution of the animal tree of life. Nature 2008; 452:745-9.
20. Ruiz-Trillo I, Roger AJ, Burger G et al. A phylogenomic investigation into the origin of metazoa. Mol Biol Evol 2008; 25:664-72.
21. Lavrov DV, Forget L, Kelly M et al. Mitochondrial genomes of two demosponges provide insights into an early stage of animal evolution. Mol Biol Evol 2005; 22:1231-9.
22. Signorovitch AY, Buss LW, Dellaporta SL. Comparative genomics of large mitochondria in placozoans. PLoS Genet 2007; 3:e13.
23. DeSalle R, Schierwater B. An even "newer" animal phylogeny. Bioessays 2008; 30:1043-7.
24. Schierwater B, DeSalle R. Can we ever identify the Urmetazoan? 10.1093/icb/icm040. Integr Comp Biol 2007; 47:670-676.
25. Schierwater B, Kamm K, Srivastava M, et al. The early ANTP gene repertoire: insights from the placozoan genome. PLoS ONE 2008; 3:e2457.
26. Yanze N, Spring J, Schmidli C et al. Conservation of Hox/ParaHox-related genes in the early development of a cnidarian. Dev Biol 2001; 236:89-98.
27. Quiquand M, Yanze N, Schmich J, et al. More constraint on ParaHox than Hox gene families in early metazoan evolution. Dev Biol 2009; 328:173-87.
28. Hui JH, Holland PW, Ferrier DE. Do cnidarians have a ParaHox cluster? Analysis of synteny around a Nematostella homeobox gene cluster. Evol Dev 2008; 10:725-30.
29. Chiori R, Jager M, Denker E, et al. Are Hox genes ancestrally involved in axial patterning? Evidence from the hydrozoan Clytia hemisphaerica (Cnidaria). PLoS ONE 2009; 4:e4231.
30. Schierwater B, Kuhn K. Homology of Hox genes and the zootype concept in early metazoan evolution. Mol Phylogenet Evol 1998; 9:375-81.
31. Cartwright P, Schierwater B, Buss LW. Expression of a Gsx parahox gene, Cnox-2, in colony ontogeny in Hydractinia (Cnidaria: Hydrozoa). J Exp Zoolog B Mol Dev Evol 2006; 306:460-9.
32. Seo HC, Edvardsen RB, Maeland AD, et al. Hox cluster disintegration with persistent anteroposterior order of expression in Oikopleura dioica. Nature 2004; 431:67-71.
33. Baguna J, Riutort M. The dawn of bilaterian animals: the case of acoelomorph flatworms. Bioessays 2004; 26:1046-57.
34. Deutsch JS. Do acoels climb up the "Scale of Beings"? Evol Dev 2008; 10:135-40.
35. Kamm K, Schierwater B. Ancient complexity of the nonHox ANTP gene complement in the anthozoan Nematostella vectensis. Implications for the evolution of the ANTP superclass. J Exp Zoolog B Mol Dev Evol 2006; 306:589-96.
36. Larroux C, Fahey B, Degnan SM, et al. The NK homeobox gene cluster predates the origin of Hox genes. Curr Biol 2007; 17:706-10.
37. Finnerty JR, Pang K, Burton P et al. Origins of bilateral symmetry: Hox and dpp expression in a sea anemone. Science 2004; 304:1335-7.

Chapter 6

Evolution of Hox Complexes

David E.K. Ferrier*

Abstract

Recent years have seen a plethora of ideas and hypotheses, and lots of debate, about the origin and evolution of the Hox gene cluster. Here I will attempt to summarize these hypotheses, identify their strengths and weaknesses and highlight the types of new data that may lead to further resolution of the competing ideas. The major theme is that Hox genes originated very early in animal evolution and extensive independent duplications occurred in major lineages. Duplications however have not been the only route to change in the composition and structure of the Hox cluster, as extensive gene losses have occurred as well. Indeed it is gene loss that is one of the main obstacles in our understanding of the origin and evolution of Hox clusters. Matters should be improved with wider taxon sampling along with a clearer understanding of how duplicated genes evolve.

Introduction

Even before the Hox genes had been cloned and the homeobox discovered, it was hypothesized by Ed Lewis that the genes of the Hox cluster, or complexes, had arisen via tandem duplication.[1-3] This was beautifully confirmed once the homeobox was discovered to be present in each homeotic gene of the *Drosophila melanogaster* Antennapedia (ANT-C) and Bithorax (BX-C) Complexes, which together constitute the Hox cluster of the fly.[4,5] The sequence similarity between the Hox genes, mainly in the homeobox, implied a homologous relationship amongst the genes.

Rapidly following on from the discovery of the homeobox in the Hox cluster it was found that the motif was widespread in many other genes, such that a typical bilaterian, such as the basal chordate lineage of amphioxus, possesses over 100 homeobox genes (133 genes in amphioxus[6]).This diversity of homeobox genes can be ordered into families, such as engrailed, even-skipped, Pax6 or Not, whereby each family is intended to designate a monophyletic group of genes represented by a single gene in the ancestor of the bilaterians (with one or two possible exceptions). These families can then be ordered into 11 classes (ANTP, PRD, ZF, TALE, CERS, POU, LIM, CUT, HNF, SINE, PROS), at least in the Bilateria,[6,7] on the basis of homeodomain sequence phylogenies aided by comparison of other motifs outside of the homeodomain. The *Drosophila Antennapedia* Hox gene lends its name to the class of homeobox gene, the Antennapedia class (ANTP-class), which contains the Hox genes along with some closely related homeobox gene families such as those constituting the ParaHox and NK genes (see below).

Origin of the ProtoHox Gene

The sequence similarities amongst homeobox genes are consistent with the families in the ANTP-class and the members of the Hox cluster, having evolved via duplications from Proto-ANTP and ProtoHox genes respectively.[8] Much more uncertainty surrounds questions such

*David E.K. Ferrier—The Scottish Oceans Institute, University of St Andrews, East Sands, St Andrews, Fife, KY16 8LB, UK. Email: dekf@st-andrews.ac.uk

Hox Genes: Studies from the 20th to the 21st Century, edited by Jean S. Deutsch.

as, when did these duplications occur and where did the ProtoHox gene come from? Such questions are plagued by the reliance upon deductions from phylogenetic trees built with homeodomain sequences, which necessarily have relatively few informative residues, and the tree topology relies on little phylogenetic signal. There may also be problems with the way we interpret gene evolution from phylogenetic trees (see below and discussed in Ferrier.[9])

Nevertheless, several models have been proposed suggesting some alternatives for the source of the ProtoHox gene. The model of Gauchat et al[10] proposes the origin of the ProtoHox gene from the Evx family, stemming from the clustering of *Evx* genes with the Hox clusters of chordates and cnidarians. An alternative proposal for the origin of the ProtoHox gene is from within the NK cluster.[11] This model arose from the observation that the sponge, *Amphimedon queenslandica*, contains ANTP-class genes related to the NK genes, but does not contain any genes with similarity to the Hox/ParaHox families. Implicit in this model is the supposition that the Porifera are the basal lineage in the animal kingdom, as supported by most phylogenetic analyses.

Dellaporta et al[12] favour an alternative phylogenetic scheme based on mtDNA analyses (that is not however supported by the largest molecular data-sets[13]), in which the phylum Placozoa, represented by *Trichoplax adhaerens*, is a more basal lineage than sponges and the *Trichoplax* Hox-like gene *Trox2* is a representative, or a relict, of the actual ProtoHox gene itself. This then requires loss of Hox-like genes in the *Amphimedon* lineage.[14,15]

A counter to both of the 'ProtoHox from NK' and 'Trox2 as ProtoHox' models is the hypothesis of Peterson and Sperling,[16] which uses topologies of phylogenetic trees to deduce extensive ANTP-class gene loss in both sponges and Placozoa. Wider sampling of sponge taxa is essential to resolve the ancestral gene complement for this phylum. Also we need a better understanding of how duplicated genes evolve and how they behave in phylogenetic trees. The Peterson and Sperling[16] interpretation follows the traditional interpretation of gene trees, assuming that the two daughters of a gene duplication will diverge to similar extents. If an alternative mode of evolution turns out to be common, as required by the NK and Trox2 models above (and called the Trox2 model in Ferrier[9]), whereby, of the two daughters from a duplication one retains more similarity to the parent gene whilst the second diverges away (see the Tandem Duplication model, Model I in Fig. 1; and Fig. 4B in ref. 9), then we would need to seriously revise our deductions from gene trees and the logic employed by Peterson and Sperling[16] would not be applicable. Indeed asymmetric evolution of duplicated genes is documented and may be associated with subfunctionalization or neofunctionalization of one of the daughter duplicates.[17] One undeniable, intriguing observation however is that *Amphimedon* develops via a larva with an anterior-posterior axis, but it does not have anything like a Hox gene.[11]

Origin of the Hox Cluster from a ProtoHox Cluster, or Not?

Number of Genes in a ProtoHox Cluster

Once the ProtoHox gene evolved, how did it give rise to the Hox gene cluster and the related genes of the ParaHox cluster? Hypotheses dealing with this question can be divided into those that involve a ProtoHox cluster versus those that don't (see Fig. 1). The ProtoHox cluster was first proposed following the discovery of the amphioxus ParaHox cluster.[18] The organisation of the genes *Gsx*, *Xlox* and *Cdx* into a gene cluster which exhibited colinearity (the order of the genes along the chromosome corresponds to the order of the expression domains along the anterior-posterior axis), along with the phylogenetic relationships of the genes relative to the Hox genes led to the model whereby the ParaHox cluster is the evolutionary sister, or paralog, to the Hox cluster. Consequently deep in animal evolution a hypothetical ProtoHox cluster existed that duplicated to produce both the Hox and ParaHox clusters contemporaneously. In the original formulation of this model the ProtoHox cluster was hypothesized to have contained four genes, with the ParaHox cluster subsequently losing a gene paralogous with the central Hox genes (see Fig. 1V).[18]

Following the initial formulation of the 4-gene ProtoHox hypothesis, various 3-gene and 2-gene ProtoHox models have been proposed.[19-23] In large part these models were stimulated by the emerging data from nonbilaterian animals (Cnidaria, Placozoa and Porifera), which have benefited in recent years from whole genome sequencing.[11,13,24,25] Much of the uncertainty about which model is most realistic again comes back to the poor resolution in homeodomain phylogenetic trees at levels deeper than the family. Although it is widely acknowledged that the resolution between homeodomain families is poor in trees, the effect that this has on relative support for the different Hox evolution models has not been statistically examined until very recently. Using Maximum Likelihood and Bayesian based statistical tests, Lanfear et al[26] now provide evidence that a 3-gene or 4-gene ProtoHox model is the most likely and various formulations of a 2-gene model are reliably rejected.

The 2-gene ProtoHox models, which have now been statistically rejected, stemmed from the hypothesis that cnidarians possess only anterior and posterior Hox and ParaHox genes and no orthologs of the Xlox/Hox3 or central Hox families.[22] Wider taxon sampling in the Cnidaria has now confirmed that cnidarians do in fact contain an *Xlox* ortholog after all.[27] There is still some uncertainty over the relationship between the *Nematostella* gene variously identified as *NvHD065* or *NvXlox/Cdx*[23,28] and the ParaHox genes of bilaterians, the former having a central place in the version of a 2-gene model proposed by Chourrout et al.[23] But with the discovery of a clearer *Xlox* ortholog in other cnidarians and the existence of a candidate *Cdx* ortholog in at least two cnidarians (*EdCnox4* from *Eleutheria dichotoma*[19,29] and *Anthox4* from *Metridium senile*[19,30])and with the widespread presence of a *Gsx* ortholog in many cnidarian species (Finnerty et al[31] and references therein), it seems reasonable to posit the existence of all three ParaHox genes in the Cnidarian-Bilaterian Ancestor (CBA).

Given the general support for the existence of ParaHox genes in cnidarians from phylogenetic trees it is somewhat surprising that the original analysis of synteny between *Nematostella* and bilaterians is consistent with the existence of a dispersed Hox cluster (Table 1), but no ParaHox locus was identified.[24] This has now been clarified by the work of Hui and colleagues.[32] The putative ParaHox cluster of *Nematostella*, containing *Anthox2/NvGsx* and *NvHD065/NvXloxCdx* on scaffold 27, which was not detected as being syntenic to the ParaHox loci of humans in a genome-wide analysis,[24] can be seen to reside in a clear, statistically significant region of synteny between *Nematostella* and humans when the scale of analysis is focused on the immediate neighbourhood of the *Nematostella* ParaHox genes rather than the entire *Nematostella* scaffold.[32] Clearly then, since a cnidarian possesses genomic loci that are syntenic and hence homologous to the Hox and ParaHox loci of bilaterians, the event that gave rise to these two loci occurred before the CBA.

Alternatives to a ProtoHox Cluster

The next question is how did these distinct Hox and ParaHox loci arise? In the ProtoHox hypotheses the evolutionary event was a whole cluster duplication (Models II-V in Fig. 1), which may have been a segmental duplication event followed by a translocation of the ParaHox cluster.[33] An alternative to the ProtoHox cluster models is the hypothesis that a Hox-like cluster was generated via a series of individual gene duplications such that it contained both Hox and ParaHox precursor genes. This precursor cluster was then split into distinct Hox and ParaHox clusters[28] (Model I in Fig. 1). This Tandem Duplication model is not well supported in the statistical tests of Lanfear et al[26] and so although it cannot be as clearly rejected as the 2-gene ProtoHox models it seems less likely than the 3- and 4-gene models.

A further alternative idea is that cnidarians do not contain Hox and ParaHox genes and that the cnidarian Hox-like sequences are the result of independent duplications from those that generated the bilaterian Hox and ParaHox genes.[34] Undoubtedly there have been some Hox-like gene duplications that could be cnidarian specific, but the extreme version of the hypothesis of independent cnidarian Hox-like duplications, in which Hox and ParaHox loci are not homologous between cnidarians and bilaterians, is now clearly untenable.[23,24,28,32] The possibility of an ancestral 2-gene cluster, whether of anterior-like and posterior-like genes,[22,34] or anterior-like and Hox3/Xlox genes,[23] is also now rejected by the work of Lanfear et al.[26]

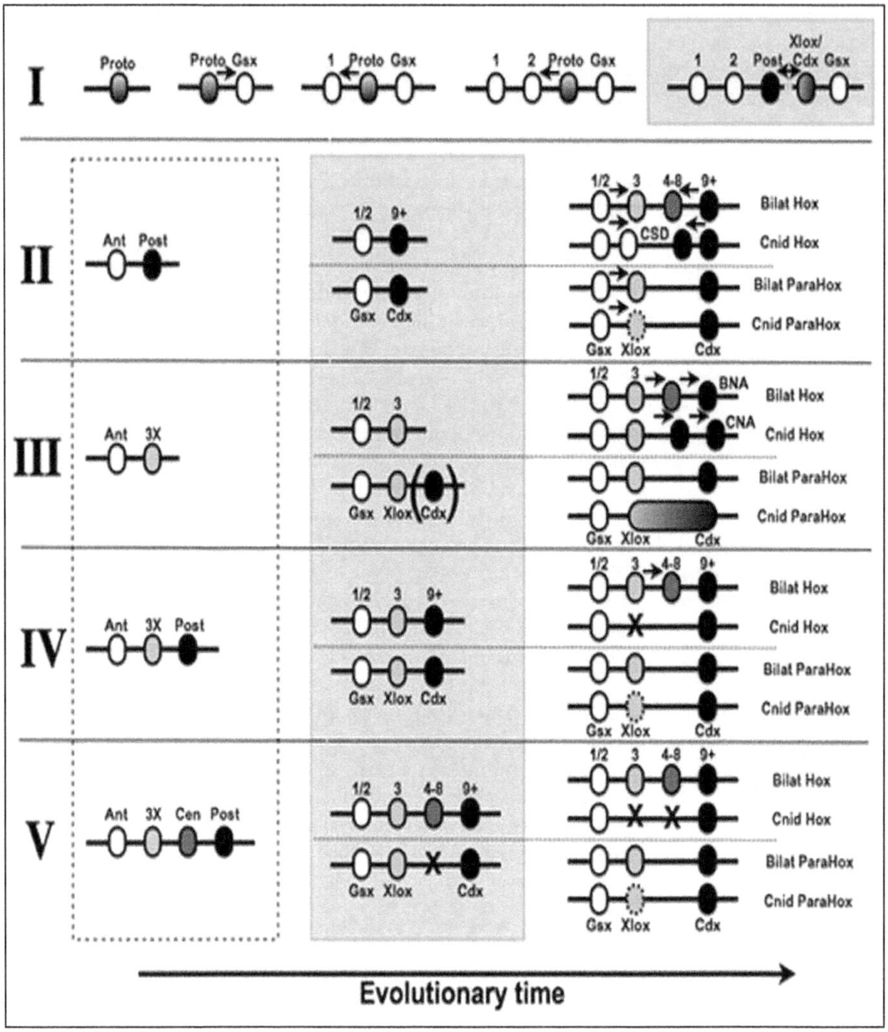

Figure 1. Summary of the alternative models proposed for the origin and evolution of the Hox (and ParaHox) clusters. All except Model I invoke a ProtoHox cluster, the different hypothesized ProtoHox clusters being enclosed in the dashed box. For each model hypothesizing a ProtoHox cluster the evolution of the Hox clusters is given above the dotted line (bilaterian = 'Bilat Hox'; cnidarian = 'Cnid Hox'), whilst the evolution of the ParaHox clusters is below the dotted line (bilaterian = 'Bilat ParaHox'; cnidarian = 'Cnid ParaHox'). Evolutionary time progresses from left to right. **Model I—Tandem Duplication** is adapted from Ryan et al[28] and hypothesizes a ProtoHox gene ('Proto') that resides in an expanding gene cluster and repeatedly duplicates to produce the precursors for the different Hox and ParaHox gene families, finally evolving into the precursors for the *Posterior Hox* and *Cdx* genes before the Precursor cluster breaks into the Hox and ParaHox clusters (broken horizontal line). Models II and III are alternative versions of a 2-gene ProtoHox. Legend continued on following page.

Figure 1, continued. **Model II—2-gene A** is adapted from Garcia-Fernàndez,[22] and requires extensive independent tandem duplications (denoted by small arrows) within the distinct Hox and ParaHox clusters after they have arisen from a ProtoHox cluster of two genes; one ProtoHox gene is the ancestor of Gsx and *Hox1/2* (= *'Ant'*) whilst the second is the ancestor for *Cdx* and the *Hox9+* genes (= *'Post'*). Within the Hox clusters the cnidarian genes other than those orthologous with *Hox1/2* and *Hox9+* are independent duplications ('CSD' = Cnidarian Specific Duplications). The dotted boundary around the cnidarian *Xlox* gene in Models II, IV and V represents the fact that *Xlox* was thought to be absent from cnidarians at the time each model was originally proposed, but has now been shown to be present in some cnidarians.[27] **Model III—2-gene B** is adapted from Chourrout et al[23] and hypothesizes a 2-gene ProtoHox cluster containing the ancestor of *Hox3* and *Xlox* (= '3X') instead of the *'Post'* ancestor of Model II.Figure 1, legend continued from previous page. This model does not distinguish whether the ParaHox cluster of the Cnidarian-Bilaterian Ancestor (CBA) contained 2 genes (*Gsx* and *Xlox*) or 3 genes (*Gsx, Xlox* and *Cdx*) (denoted by the brackets around the CBA *Cdx* gene). In the latter case the present-day cnidarian ParaHox cluster (represented by *Nematostella vectensis*) has been reduced back to a 2-gene cluster, with a gene of indeterminate orthology between *Xlox* and *Cdx* (denoted by the stretched gene symbol). Extensive independent duplications are hypothesized for the generation of the Bilaterian Non-Anterior genes ('BNA') and the Cnidarian Non-Anterior genes ('CNA'). **Model IV—3-gene** adapted from Finnerty and Martindale[19] and Ferrier and Holland,[20] in which the central Hox genes ('4-8') evolved within the bilaterian Hox cluster and the cnidarian lineage lost a *Hox3* ortholog. **Model V—4-gene** adapted from Ferrier and Holland[20] involves loss of *Hox3* and *Hox4-8* orthologs from the cnidarian Hox cluster and loss of a ParaHox gene paralogous to *Hox4-8* in the CBA. The shaded box highlights the hypothesized organization of the Hox/ParaHox genes in the CBA for each model. Small arrows within clusters denote duplication events. Although these are given as arrows the actual direction of the duplication is often unknown (i.e., whether the central Hox might have duplicated from either a *Posterior Hox* or a *Hox3* ancestor). Gene loss events are denoted as 'X' on the horizontal lines, which themselves denote the chromosome.

Any remaining controversy over whether cnidarians possess Hox (and ParaHox) genes stems from the ambiguity of the phrase 'true' or 'definitive' Hox genes as used in Kamm et al,[34] implying an understanding of the fundamental function of Hox genes and their organization, and this function and organization being different between the Cnidaria and Bilateria. This functional and organizational information is usually lacking for the genes sampled from most taxa and it is not always clear how expression should be related to possible roles in axial development and how variable Hox gene expression can be (discussed in Amemiya and Wagner[35]). Further difficulties arise from the ambiguities in ortholog identification for the cnidarian genes, even between different species of cnidarian, which in turn can confound comparisons when gene loss has occurred (see below). Also it is far from clear how to compare developmental stages and even embryonic axes between the earliest branches in the animal kingdom (e.g., cnidarian planula or polyps with bilaterian embryos and larvae[36-38]). In addition it is now apparent that the organization of the Hox genes in the Bilateria is subject to extensive rearrangement (reviewed in Monteiro and Ferrier[39]). So lack of clustering is not a characteristic that can be used to distinguish whether a cnidarian gene is a Hox gene or not. Genomic context and synteny can sometimes still be informative and it is intriguing that 3 of the 4 *Nematostella* scaffolds bearing putative Hox genes (judging from their phylogenetic affinities) show significant synteny with the human Hox chromosomes[24,28] (see Table 1), which implies *Nematostella* is another example of an animal with a dispersed Hox cluster. The cnidarian lineage clearly did not evolve before a cluster of at least three genes homologous to the bilaterian Hox cluster existed (contra Kamm et al[34]).

A classification of Hox and nonHox genes based upon sequence similarity and genomic context is thus our best method of comparison at present. On the basis of phylogenetic trees and synteny it is clear that cnidarians do possess both Hox and ParaHox genes. The question remains open however, as to what the function of these genes is in many extant lineages, such as the Cnidaria and Placozoa, and then what the function was in the extinct ancestors of these lineages and the bilaterians? Wider taxon sampling from the nonbilaterian phyla should enable clearer pictures of

Table 1. Synteny mapping between Nematostella and humans implies a dispersed Hox gene cluster in this cnidarian. Other Nematostella scaffolds with significant synteny to human Hox chromosomes are scaffolds 53, 46 and 5.[24]

Nematostella Hox-Like Genes	Bilaterian Affinity (from Ryan et al[28])	JGI Scaffold Number	Significant Synteny to Human Hox Chromosomes (from Putnam et al[24])
Anthox6, 8a, 8b, 7 (& *Evx*)	*Hox1/2* (& *Evx*)	61	–
Anthox6a	*Hox1*	26	–
Anthox1a (& *9*)	*Posterior Hox* (& *Gsx* or *Mox*)	3	–
Anthox1	*Posterior Hox*	4	+
Anthox2/NvGsx, NvHD065/NvXlox/Cdx	ParaHox	27	+ (but synteny with ParaHox[32]),

gene loss and orthology to be constructed for the Cnidaria and the relationships to the bilaterian genes. Then careful comparisons between the function of truly orthologous genes can be performed to assess whether the function of Hox genes as generally (but perhaps still poorly) understood from bilaterians also applies to the nonbilaterian genes. It is clear that the expression of the Hox genes in Cnidaria is regionalized during embryo and larval development,[28,40] which is consistent with some sort of role in axial patterning, although Kamm et al[34] disagree. What effects the nonbilaterian Hox genes have on cell fates and axial patterning is still not well understood and it may well be difficult to assay whether any kind of homeotic phenotype, such as those produced by Hox gene perturbation in bilaterians, has occurred in a cnidarian. Intriguingly the only Hox/ParaHox gene in Placozoa, *Trox2*, is involved in growth and fission and so potentially is involved in cell division and differentiation.[14] Elucidation of the function of further developmental genes in *Trichoplax* will help to interpret this *Trox2* phenotype and facilitate comparisons with other phyla. Ideally, in the future, such assays will be extended back into embryogenesis once a means to obtain the embryos reliably is found.[13] With a clearer understanding of gene relationships in hand, more extensive functional analyses in several cnidarians will then also be essential.[41]

Expansion and Contraction of the Number of Hox Genes in Evolution

The general scheme for Hox evolution during animal evolution is one of expansion by gene duplication,[8,42-45] which perhaps harks back to the earliest views on Hox involvement in body plan evolution that posited, in their strongest form, an increase in Hox gene number with the increasing complexity of animal body plans.[3,46] This has obviously been significantly moderated as an extensive complement of Hox genes were found across the Bilateria, but a connection between Hox gene number and animal evolution still seems intuitively attractive.[21,47]

There are however many clear examples of Hox gene loss in animal evolution. When a fourteenth Hox gene was first found in amphioxus it was thought that this might be an amphioxus-specific duplication[48] (perhaps analogous to some of the cnidarian Hox-like genes discussed above), until *Hox 14* genes were discovered in various craniate taxa.[49,50] The likelihood is now that the chordate ancestor possessed a *Hox14* gene, and certainly the craniate ancestor did.[50,51] Craniate evolution has subsequently involved loss of *Hox14* in several independent lineages (e.g., teleosts and tetrapods). Interestingly a *Hox15* gene has now been discovered in amphioxus[52] and the possibility raised that other deuterostomes may also possess a *Hox15* that has been very widely lost in many lineages.

Table 2. Increasing number of ANTP-class and Hox/ParaHox genes in higher lineages of animals. The gene numbers have been deduced from whole genome sequences.

	ANTP-class	Hox + ParaHox	Reference
Choanoflagellate (*Monosiga brevicollis* (= nonmetazoan, sister group and outgroup to animals)	0	0	King et al[61]
Nonbilateria Porifera (*Amphimedon queenslandica*)	8	0	Larroux et al[11,25]
Placozoa (*Trichoplax adhaerens*)	14	1(+1?)	Schierwater et al[15]; Srivastava et al[13]
Cnidaria (*Nematostella vectensis*)	72-78	10	Chourrout et al[23]; Ryan et al[28,62]
Bilateria Arthropoda (*Tribolium castaneum*)	45	14	Richards et al[63]
Chordata (*Branchiostoma floridae*)	60	18	Takatori et al[6]; Holland et al[52]
Chordata (*Homo sapiens*)	100	45	Holland et al[7]

The duplicated Hox clusters of vertebrates, and teleost fish in particular, provide a host of examples of Hox gene loss.[53] However, cluster duplication is not absolutely necessary before Hox genes can be lost. Sea urchins have lost a *Hox4* gene[54] and urochordates have significantly reduced their complement of Hox genes from the ancestral chordate condition.[55,56] Outside of the deuterostomes the lophotrochozoan phylum of Platyhelminthes contains a clear example of Hox gene loss. The parasitic flatworm *Schistosoma mansonii* now only possesses 5 Hox genes, a reduction from the ancestral number of 7-11 for the Lophotrochozoa.[57,58] In the ecdysozoan clade the nematode lineage leading to *Caenorhabditis elegans* is yet another clear example of Hox gene loss.[59]

This indicates that we must err on the side of caution when hypothesizing expansion of Hox gene numbers during animal evolution. This is particularly true when we are restricted to sampling only a few lineages from entire phyla, ideally at the whole genome level (e.g., *Amphimedon, Trichoplax* and *Nematostella*), as gene loss could well be a significant factor,[16] notwithstanding the caveat about interpreting tree topology to deduce gene loss (see above). Although the picture is reassuringly consistent at present, with a gradual increase in the numbers of ANTP-class and then Hox genes at successively higher nodes in animal evolution (i.e., the origin of animals, the origin of Bilateria, then the origin of vertebrates) (Table 2), it is intriguing that the gene families present across some basal phyla, for example Placozoa and Porifera, do not tend to overlap, which would be consistent with gene loss.[16] What is clearly required is more extensive taxon sampling, ideally via whole genome sequencing to reduce the chances of missing genes (now that new sequencing technologies are bringing this into the realms of possibility). Although no-one seriously doubts that Hox genes originated within the animal kingdom and there must inevitably have been some period of time over which Hox gene number increased, we still have plenty to discover about exactly when they arose, when they expanded in number and how the Hox cluster as a whole originated in animal evolution.

Conclusion

Much is still to be determined about the axial patterning role of Hox genes and whether it is homologous across the animals. Certainly at least one sponge seems to develop an anterior-posterior embryonic axis without Hox genes, challenging any views of Hox patterning being integral to the development of all animals.[11,60] The weight of evidence now supports the origin of the Hox cluster by a whole-cluster duplication from a 3- or 4-gene ProtoHox cluster before the origin of the Cnidaria. But these models of evolution will need to be constantly revisited as new taxa are examined, new genes discovered and our gene phylogenies refined, with a clearer appreciation for the prevalence of gene loss and the utility of analyzing the genomic neighbourhood of a gene when tracing its evolutionary history.

References

1. Lewis EB. Pseudoallelism and gene evolution. Cold Spring Harb Symp Quant Biol 1951; 16:159-174.
2. Lewis EB. Genes and developmental pathways. American Zoologist 1963; 3:33-56.
3. Lewis EB. A gene complex controlling segmentation in Drosophila. Nature 1978; 276(5688):565-570.
4. McGinnis W, Garber RL, Wirz J et al. A homologous protein-coding sequence in Drosophila homeotic genes and its conservation in other metazoans. Cell 1984; 37(2):403-408.
5. McGinnis W, Levine MS, Hafen E et al. A conserved DNA sequence in homoeotic genes of the Drosophila Antennapedia and bithorax complexes. Nature 1984; 308(5958):428-433.
6. Takatori N, Butts T, Candiani S et al. Comprehensive survey and classification of homeobox genes in the genome of amphioxus, Branchiostoma floridae. Dev Genes Evol 2008; 218(11-12):579-590.
7. Holland PWH, Booth HA, Bruford EA. Classification and nomenclature of all human homeobox genes. BMC Biol 2007; 5:47.
8. Garcia-Fernàndez J. The genesis and evolution of homeobox gene clusters. Nat Rev Genet 2005; 6(12):881-892.
9. Ferrier DEK. Evolution of Hox gene clusters. In: Papageorgiou S, ed. HOX gene expression. First ed. Austin, Texas, and New York: Landes Bioscience/Eurekah.com and Springer Science 2007; 53-67.
10. Gauchat D, Mazet F, Berney C et al. Evolution of Antp-class genes and differential expression of Hydra Hox/paraHox genes in anterior patterning. Proc Natl Acad Sci USA 2000; 97(9):4493-4498.
11. Larroux C, Fahey B, Degnan SM et al. The NK homeobox gene cluster predates the origin of Hox genes. Curr Biol 2007; 17(8):706-710.
12. Dellaporta SL, Xu A, Sagasser S et al. Mitochondrial genome of Trichoplax adhaerens supports Placozoa as the basal lower metazoan phylum. Proc Natl Acad Sci USA 2006; 103(23):8751-8756.
13. Srivastava M, Begovic E, Chapman J et al. The Trichoplax genome and the nature of placozoans. Nature 2008; 454(7207):955-960.
14. Jakob W, Sagasser S, Dellaporta S et al. The Trox-2 Hox/ParaHox gene of Trichoplax (Placozoa) marks an epithelial boundary. Dev Genes Evol 2004; 214(4):170-175.
15. Schierwater B, Kamm K, Srivastava M et al. The early ANTP gene repertoire: insights from the placozoan genome. PLoS ONE 2008; 3(8):e2457.
16. Peterson KJ, Sperling EA. Poriferan ANTP genes: primitively simple or secondarily reduced? Evol Dev 2007; 9(5):405-408.
17. Clément Y, Tavares R, Marais GA. Does lack of recombination enhance asymmetric evolution among duplicate genes? Insights from the Drosophila melanogaster genome. Gene 2006; 385:89-95.
18. Brooke NM, Garcia-Fernàndez J, Holland PWH. The ParaHox gene cluster is an evolutionary sister of the Hox gene cluster. Nature 1998; 392(6679):920-922.
19. Finnerty JR, Martindale MQ. Ancient origins of axial patterning genes: Hox genes and ParaHox genes in the Cnidaria. Evol Dev 1999; 1(1):16-23.
20. Ferrier DEK, Holland PWH. Ancient origin of the Hox gene cluster. Nat Rev Genet 2001; 2(1):33-38.
21. Baguña J, Riutort M. The dawn of bilaterian animals: the case of acoelomorph flatworms. Bioessays 2004; 26(10):1046-1057.
22. Garcia-Fernàndez J. Hox, ParaHox, ProtoHox: facts and guesses. Heredity 2005; 94(2):145-152.
23. Chourrout D, Delsuc F, Chourrout P et al. Minimal ProtoHox cluster inferred from bilaterian and cnidarian Hox complements. Nature 2006; 442(7103):684-687.
24. Putnam NH, Srivastava M, Hellsten U et al. Sea anemone genome reveals ancestral eumetazoan gene repertoire and genomic organization. Science 2007; 317(5834):86-94.
25. Larroux C, Luke GN, Koopman P et al. Genesis and expansion of metazoan transcription factor gene classes. Mol Biol Evol 2008; 25(5):980-996.
26. Lanfear R, Bromham L. Statistical tests between competing hypotheses of Hox cluster evolution. Systematic Biology 2008; 57(5):708-718.

27. Quiquand M, Yanze N, Schmich J et al. More constraint on ParaHox than Hox gene families in early metazoan evolution. Developmental Biology 2009; 328:173-187.
28. Ryan JF, Mazza ME, Pang K et al. Pre-bilaterian origins of the Hox cluster and the Hox code: evidence from the sea anemone, Nematostella vectensis. PLoS ONE 2007; 2(1):e153.
29. Kuhn K, Streit B, Schierwater B. Homeobox genes in the cnidarian Eleutheria dichotoma: evolutionary implications for the origin of Antennapedia-class (HOM/Hox) genes. Mol Phylogenet Evol 1996; 6(1):30-38.
30. Finnerty JR, Martindale MQ. Homeoboxes in sea anemones (Cnidaria:Anthozoa): a PCR-based survey of Nematostella vectensis and Metridium senile. Biol Bull 1997; 193(1):62-76.
31. Finnerty JR, Paulson D, Burton P et al. Early evolution of a homeobox gene: the parahox gene Gsx in the Cnidaria and the Bilateria. Evol Dev 2003; 5(4):331-345.
32. Hui JHL, Holland PWH, Ferrier DEK. Do cnidarians have a ParaHox cluster? Analysis of synteny around a Nematostella homeobox gene cluster. Evol Dev 2008; 10(6):725-730.
33. Minguillón C, Garcia-Fernàndez J. Genesis and evolution of the Evx and Mox genes and the extended Hox and ParaHox gene clusters. Genome Biol 2003; 4(2):R12.
34. Kamm K, Schierwater B, Jakob W et al. Axial patterning and diversification in the Cnidaria predate the Hox system. Curr Biol 2006; 16(9):920-926.
35. Amemiya CT, Wagner GP. Animal evolution: when did the 'Hox system' arise? Curr Biol 2006; 16(14):R546-548.
36. Meinhardt H. The radial-symmetric hydra and the evolution of the bilateral body plan: an old body became a young brain. Bioessays 2002; 24(2):185-191.
37. Baguña J, Martinez P, Paps J et al. Back in time: a new systematic proposal for the Bilateria. Philos Trans R Soc Lond B Biol Sci 2008; 363(1496):1481-1491.
38. Hejnol A, Martindale MQ. Acoel development supports a simple planula-like urbilaterian. Philos Trans R Soc Lond B Biol Sci 2008; 363(1496):1493-1501.
39. Monteiro AS, Ferrier DEK. Hox genes are not always Colinear. Int J Biol Sci 2006; 2(3):95-103.
40. Finnerty JR, Pang K, Burton P et al. Origins of bilateral symmetry: Hox and dpp expression in a sea anemone. Science 2004; 304(5675):1335-1337.
41. Jakob W, Schierwater B. Changing hydrozoan bauplans by silencing Hox-like genes. PLoS ONE 2007; 2(1):e694.
42. Zhang J, Nei M. Evolution of Antennapedia-class homeobox genes. Genetics 1996; 142(1):295-303.
43. Gehring WJ, Affolter M, Bürglin T. Homeodomain proteins. Annu Rev Biochem 1994; 63:487-526.
44. Schubert FR, Nieselt-Struwe K, Gruss P. The Antennapedia-type homeobox genes have evolved from three precursors separated early in metazoan evolution. Proc Natl Acad Sci USA 1993; 90(1):143-147.
45. Deutsch JS, Lopez P. Are transposition events at the origin of the bilaterian Hox complexes? In: Minelli A, Fusco G, eds. Evolving Pathways: Key themes in evolutionary developmental biology. First ed. Cambridge: Cambridge University Press, 2008.
46. Akam ME, Averof M, Castelli-Gair J et al. The evolving role of Hox genes in arthropods. Dev Suppl 1994;209-215.
47. Jiménez-Guri E, Paps J, Garcia-Fernàndez J et al. Hox and ParaHox genes in Nemertodermatida, a basal bilaterian clade. Int J Dev Biol 2006; 50(8):675-679.
48. Ferrier DEK, Minguillón C, Holland PWH et al. The amphioxus Hox cluster: deuterostome posterior flexibility and Hox14. Evol Dev 2000; 2(5):284-293.
49. Powers TP, Amemiya CT. Evidence for a Hox14 paralog group in vertebrates. Curr Biol 2004; 14(5):R183-184.
50. Kuraku S, Takio Y, Tamura K et al. Noncanonical role of Hox14 revealed by its expression patterns in lamprey and shark. Proc Natl Acad Sci USA 2008; 105(18):6679-6683.
51. Ferrier DEK. Hox genes: Did the vertebrate ancestor have a Hox14? Curr Biol 2004; 14(5):R210-211.
52. Holland LZ, Albalat R, Azumi K et al. The amphioxus genome illuminates vertebrate origins and cephalochordate biology. Genome Res 2008; 18(7):1100-1111.
53. Meyer A, Malaga-Trillo E. Vertebrate genomics: More fishy tales about Hox genes. Curr Biol 1999; 9(6):R210-213.
54. Long S, Martinez P, Chen WC et al. Evolution of echinoderms may not have required modification of the ancestral deuterostome HOX gene cluster: first report of PG4 and PG5 Hox orthologues in echinoderms. Dev Genes Evol 2003; 213(11):573-576.
55. Ikuta T, Yoshida N, Satoh N et al. Ciona intestinalis Hox gene cluster: Its dispersed structure and residual colinear expression in development. Proc Natl Acad Sci USA 2004; 101(42):15118-15123.
56. Seo HC, Edvardsen RB, Maeland AD et al. Hox cluster disintegration with persistent anteroposterior order of expression in Oikopleura dioica. Nature 2004; 431(7004):67-71.
57. Pierce RJ, Wu W, Hirai H et al. Evidence for a dispersed Hox gene cluster in the platyhelminth parasite Schistosoma mansoni. Mol Biol Evol 2005; 22(12):2491-2503.

58. Balavoine G, de Rosa R, Adoutte A. Hox clusters and bilaterian phylogeny. Mol Phylogenet Evol 2002; 24(3):366-373.
59. Aboobaker AA, Blaxter ML. Hox gene loss during dynamic evolution of the nematode cluster. Curr Biol 2003; 13(1):37-40.
60. Slack JM, Holland PWH, Graham CF. The zootype and the phylotypic stage. Nature 1993; 361(6412):490-492.
61. King N, Westbrook MJ, Young SL et al. The genome of the choanoflagellate Monosiga brevicollis and the origin of metazoans. Nature 2008; 451(7180):783-788.
62. Ryan JF, Burton PM, Mazza ME et al. The cnidarian-bilaterian ancestor possessed at least 56 homeoboxes: evidence from the starlet sea anemone, Nematostella vectensis. Genome Biol 2006; 7(7):R64.
63. Richards S, Gibbs RA, Weinstock GM et al. The genome of the model beetle and pest Tribolium castaneum. Nature 2008; 452(7190):949-955.

CHAPTER 7

The Nematode Story: Hox Gene Loss and Rapid Evolution

Aziz Aboobaker* and Mark Blaxter

Abstract

The loss in some taxa of conserved developmental control genes that are present in the vast majority of animal lineages is an understudied phenomenon. It is likely that in those lineages in which loss has occurred it may be a strong signal of the mode, tempo and direction of developmental evolution and thus identify ways of generating morphological novelties. Intuitively we might expect these novelties to be particularly those associated with morphological simplifications. One striking example of this has occurred within the nematodes. It appears that over half the ancestral bilaterian Hox cluster has been lost from the model organism *Caenorhabditis elegans* and its closest related species. Studying the Hox gene complement of nematodes across the phylum has shown that many, if not all these losses occurred within the phylum. Other nematode clades only distantly related to *C. elegans* have additional Hox genes orthologous to those present in the ancestral bilaterian but absent from the model nematode. In some of these cases rapid sequence evolution of the homeodomain itself obscures orthology assignment until comparison is made with sequences from multiple nematode clades with slower evolving Hox genes. Across the phylum the homeodomains of the Hox genes that are present are evolving very rapidly. In one particular case the genomic arrangement of two homeodomains suggests a mechanism for gene loss. Studying the function in nematodes of the Hox genes absent from *C. elegans* awaits further research and the establishment of new nematode models. However, what we do know about Hox gene functions suggests that the genetic circuits within which Hox genes act have changed significantly within *C. elegans* and its close relatives.

Introduction: Hox Gene Loss, the Third Way

For over twenty years Hox genes have been a Rosetta stone for our continued attempts to understand the molecular evolutionary bases of the amazing diversity of extant and extinct animal body plans.[1] A simplistic summary of what the "evo-devo" field has found would be to conclude that body plan evolution involves three major broadly defined mechanisms and in each case Hox genes provide prime examples. Firstly, we observe the redeployment of conserved developmental genetic circuits to new contexts through changes in the time and place at which they act and in the partners with which they interact during development. Secondly, we observe the invention of new genes with subsequent incorporation into existing or duplicated, paralagous networks. This has often occurred by mixing existing domains into new combinations with new activities. Alternatively duplications of existing genes, clusters of paralogous genes or of the whole genome can provide the substrate for developmental evolution. Finally and in our opinion currently

*Corresponding Author: Aziz Aboobaker—Institute of Genetics, The University of Nottingham, Queen's Medical Centre, Nottingham NG7 2UH, UK.
Email: aziz.aboobaker@nottingham.ac.uk

Hox Genes: Studies from the 20th to the 21st Century, edited by Jean S. Deutsch.

understudied, is the loss of otherwise highly conserved genes from developmental programmes in some groups of organisms. The study of Hox genes continues to provide a substrate for investigation of these three main themes as well as their interplay with each other. More than two decades after the first molecular description of the homeobox[2,3] there are still major unanswered questions to which continued research on Hox genes will make a major contribution.

In this chapter we consider Hox genes with respect to the last of these broadly defined mechanisms: gene loss. While gene loss can be inferred in all animal lineages, in some it seems to be extreme and thus might possibly be a defining mechanism for within-phylum developmental evolution. With respect to Hox genes this is perhaps most apparent within the phylum Nematoda.[4] Here we review the trajectory of ideas that have seen our view of the nematode Hox cluster change from that of a simple, ancestral (plesiomorphic) genomic structure[5,6] to a highly derived, rapidly evolving genomic structure.[4] We describe the research landmarks, such as the complete sequencing of the genome of the nematode model *Caenorhabditis elegans*,[7] which were responsible for this radical transition. In this discussion we will also highlight the need for an accurate picture of phylogenetic inter-relationships at all systematic levels when using molecular developmental data to form evolutionary developmental hypotheses about the direction, mode and tempo of evolutionary change.

The state of gene loss within the model system *C. elegans* may be a strong clue as to the nature of the molecular evolutionary events responsible for morphological and developmental evolution within the nematode lineage. Research across the whole phylum illustrates that Hox gene loss observed with *C. elegans* and its close sister taxa is indeed the most extreme state and that other nematodes retain Hox genes that have been lost in the lineage leading to *C. elegans*. We present analysis of the Hox cluster from other nematodes and show that while they have more Hox genes they still have impoverished clusters. These data also suggest interesting scenarios by which Hox genes were lost within nematodes. This in turn has implications for explaining how the loss of conserved developmental genes might occur at the molecular level and how this might be reflected in development.[4,8]

We also consider the changing developmental functions of nematode Hox genes, which still provide some of the clearest examples of the functional evolution of conserved developmental genes and how the genetic circuits they are in can change quite radically.[9-12] However, despite all this evolutionary change it also appears that some ancient aspects of Hox gene activity and interactions are conserved even in *C. elegans*.[13,14] This highlights that it is worth remembering that despite all the changes that have been rung Hox genes in nematodes are still conserved and are still specifying fates, albeit for the most part within the context of a cell lineage-based developmental mode.

In the final part of this chapter we consider the unanswered questions concerning Hox genes within the Nematoda and suggest what might be done to answer them.

The *Caenorhabditis elegans* Hox Cluster, an Extreme Case of Gene Loss

In the early 1990's work on the two major invertebrate model systems, *C. elegans* and *Drosophila melanogaster* continued to expand as additional molecular genetic technologies became available. Both models provided different strengths and advantages and the then-current view of their phylogenetic relationships to each other and vertebrates further justified their use. Both classical morphological observations and pioneering molecular phylogenetic work[15] suggested that *C. elegans* and other nematodes form an out-group to *D. melanogaster* and vertebrates. This hypothesis has been referred to as the Coelomata hypothesis as it groups those animals that form a true body cavity (coelom). From this assumption it also followed that cellular and developmental mechanisms common to the invertebrate systems would also be likely to be conserved in vertebrates.

This view also greatly influenced the interpretation of the initial discovery of Hox genes in *C. elegans*.[5,6,16] Initially four bona-fide Hox genes were discovered. They were also found to be loosely clustered and this region was among the first to be fully sequenced as part of the *C. elegans* genome

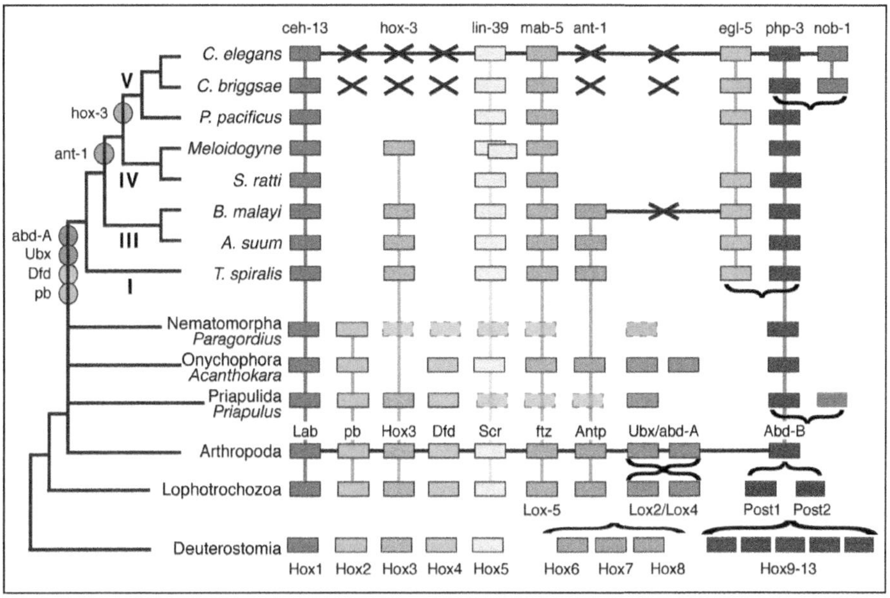

Figure 1. Hox gene evolution in the phylum Nematoda. This figure (adapted from Aboobaker and Blaxter (2003)) illustrates the pattern of Hox gene retention in the Phylum Nematoda. On the left is a schematic phylogenetic tree showing the relationships within the Nematoda (and the major clades, *sensu* Blaxter et al (1998), to which the species belong) and the relationships of other animal phyla and superphyla. On the Nematoda portion of the tree, circles indicate the latest inferred time of loss of particular Hox genes. The structure within the Ecdysozoa is here left unresolved, though Nematomorpha are believed to be closest to Nematoda and Priapulida are believed to arise basally compared to the other phyla figured. Each colored box represents a cloned and sequenced Hox gene or Hox homeodomain fragment, aligned vertically to illustrate inferred orthology relationships. Curled brackets indicate independent expansions of Hox genes of particular classes. For genome-sequenced species, the synteny relationships of Hox genes are indicated by the horizontal black lines and Hox orthologs for which there is definitive evidence of loss are indicated with heavy X marks. Within Nematoda, the two boxes placed for *Meloidogyne Scr/Hox5* indicate two very similar homeodomains cloned from *M. javanica*. The *M. incognita* genome sequence revealed that this species is a hybrid and thus we hypothesize that these homeodomains may similarly simply reflect hybrid species origin. In Nematomorpha and Priapulida, the ortholog group membership of central-group homeodomain fragments remains uncertain, indicated by the dotted outlines.

project.[17] These genes were representatives of the anterior Hox1 class (*ceh-13*), two central class genes (*lin-39* and *mab-5*) and posterior class gene (*egl-5*). In addition an *eve* class gene, in the Antennapedia superfamily, was also found within the bounds of this loose cluster. The finding that the genes sequences of this 'simple' four-gene cluster were for the most part more divergent from fly and vertebrate sequences then these groups were from each other also supported the existing systematic view.

The molecular data supporting the Coelomata hypothesis consisted of either a few gene sequences from experimentally important taxa,[15] or broader gene sampling from very few taxa.[18] On broader taxon sampling with much larger data sets of orthologous genes our view of animal phylogeny has changed drastically (Fig. 1). Landmark work by Aguinaldo et al and Halanych et al instead suggested that bilaterian animals comprise three major lineages and, significantly, that nematodes and arthropods are both members of one of these major lineages, the Ecdysozoa.[19,20]

These analyses used broader phylogenetic sampling and also selected molecular data that gave rise to shorter branch lengths. Studies that placed nematodes basal to flies and vertebrates appear to have suffered from long-branch attraction caused by rapid molecular evolution, particularly within the model taxa *C. elegans* and *D. melanogaster*. Although other recent molecular analyses have continued to propose other phylogenetic hypotheses for Metazoa,[21,22] for the most part the Ecdysozoa and Lophotrochozoa split in the protostomes is now widely accepted.[23,24]

This major upheaval of animal systematics necessitated an ongoing reinterpretation of the evolutionary trajectory of all the developmental characteristics mapped in the arthropod, nematode and other major phyla. Chief among these for the Nematoda was a need to analyze the evolutionary and functional dynamics of the Hox cluster. Had many Hox genes really been lost and if so when? What was the situation in other ecdysozoans? Were the Hox clusters ostensibly containing orthologs of all vertebrate Hox genes observed in some protostome clades (i.e., arthropods and brachiopods) indicative of the plesiomorphic state or were they the results of homoplastic independent duplications of some paralogous groups in each lineage?

The completion of the *C. elegans* genome provided a partial answer to the questions concerning nematode Hox genes. One of the last clones to be sequenced, mapping to right hand arm of chromosome III contained two tandemly arranged Hox genes, both belonging to the posterior class.[25] One of these genes, *php-3* (*posterior hox protein 3*), appeared to be much more similar to the Drosophila posterior Hox gene *Abd-B* than the others. Interestingly, the second newly discovered gene, *nob-1* (*no-back-end 1*) seemed to have a major developmental role whereas that of *php-3* appears to be only minor. Thus *C. elegans* has genes from the Hox1 (*ceh-13*), Hox5 (*lin-39*), Hox6-8 (*mab-5*) and posterior group Hox9-13 (*egl-5, php-3, nob-1*) classes. A cross-phylum collaboration between three laboratories was able to clearly establish that a sophisticated Hox cluster (in terms of membership) existed in the ancestor of all protostomes and deuterostomes containing at least 9 Hox genes representing orthologs of the *Hox1, Hox2, Hox3, Hox4, Hox5, Hox6, Hox7, Hox8* and *Hox9-13* genes in vertebrates.[26] Particularly useful in these analyses were the existence of conserved orthology group specific peptides outside of the homeodomain. The *C. elegans* cluster was shoehorned into this system with difficulty.

Significantly, this confirmed that somewhere in the evolutionary lineage leading to the nematode *C. elegans* at least five Hox orthology groups were lost. Had these losses occurred recently or were they spread out over a long evolutionary time? Can they tell us anything about how conserved genes are lost and how this might be reflected in extant developmental mechanisms?

Tracing Hox Gene Loss through the Nematode Phylum: Mode and Tempo

In order to understand Hox gene evolution in nematodes we set out to clone the Hox gene complements from representative species from across the phylum.[8] We were able to make informed choices about the taxa we chose to study because of the recent establishment of a robust molecular phylogeny for the whole phylum Nematoda.[27] This framework, based on small subunit ribosomal RNA genes, broadly splits the Nematoda into 5 major Clades (I-V) with *C. elegans* and many of the other free-living soil nematodes in Clade V. This framework has been subsequently affirmed and refined using additional taxa and genes.[28] We chose a range of species that spanned the breadth of the phylum in order to understand when the Hox gene losses and the rapid sequence evolution observed in *C. elegans* had occurred.

We used an extensive degenerate PCR screen with primers designed to the first and third helix of the homeodomain. As we cloned more Hox genes from different species we were able to iteratively redesign our primer sets to amplify divergent nematode homeodomains that might be missed by the cross-phylum primers used elsewhere.[26] With this approach we cloned Hox genes from six different species and compared them to those of *C. elegans* and those of other phyla (Fig. 1).

We found that at least two Hox genes, a *Hox3* gene and a central class gene, had been lost during the evolution of the nematode lineage. The *Hox3* gene cloned from two Clade III parasitic nematodes *Ascaris suum* and *Brugia malayi* was highly divergent. Only on cloning a slower evolving *Hox3* from the Clade I nematode *Trichinella spiralis* could we be sure that these genes were indeed *Hox3* orthologs. It may be a general feature of the *T. spiralis* genome (and possibly other Clade I nematodes) that coding sequences have evolved more slowly than other nematodes since the derivation of the phylum from the protostome ancestor. In turn this analysis allowed us to design nematode-specific *Hox3* primers with which we rescreened the species in our study. We found that there were no *Hox3* genes in any of the Clade V species surveyed. Subsequent whole-genome sequencing of *Pristionchus pacificus* has verified this absence. However, we were able to find a very divergent *Hox3* gene in the plant parasite *Meloidogyne javanica*. Without the sequences of the three other nematode *Hox3* homeodomains it would have been very difficult to correctly assign the *M. javanica* sequence to this group. This places the event of *Hox3* loss in the lineage leading to Clade V. It also suggests that, for this Hox ortholog group, very rapid sequence evolution is likely to have preceded loss and so it remains possible that any true ortholog within Clade V might not be identifiable by sequence similarity methods.

Sea Squirts and Nematodes: Why Do Both Groups Lose Hox Genes

The other metazoan group wherein there is definitive evidence of Hox gene loss and disintegration of the Hox cluster, is the Urochordata, the sea squirts or tunicates. Sea squirts are chordate relatives where, in most species, the notochord is only present during the dispersing motile 'tadpole larva' phase. The larva has a body plan reminiscent of models of the early chordate, with a cephalic capsule and anterior mouth, a muscular elongated body supported by a notochord, a dorsal nerve cord and metamerically repeated muscles. The adults are mostly sessile or nonmotile, a habit achieved after a remarkable metamorphosis.

Two urochordates have been examined in detail for Hox gene complement. *Ciona intestinalis* has sessile adults and the tadpole larva and has nine Hox genes;[29] there are no *Hox-7*, *-8*, *-9* or *-11* homologs in *Ciona*. Anterior-posterior expression patterns of these genes along the larval body axis is as expected from other Metazoa. The genes are located on two chromosomes, with *Ci-Hox1, 2, 3, 4, 5, 6* and *10* on one chromosome and *Ci-Hox12* and *13* on another.[29] Within the chromosome containing the bulk of the Hox genes, the order is rearranged compared to that expected from anterior-posterior expression, with *Hox10* found between *Hox2, 3* and *4* and *Hox5* and *6*. *Ci-Hox1* is separated from the other Hox genes on the same chromosome by about 50% of the chromosome length.[29]

In the urochordate class Appendicularia, the tadpole larva has neotenously become the fertile and motile adult. Phylogenetic reconstruction suggests that this is through loss of the sessile stage. In the genome of the appendicularian *Oikopleura dioica*, there are also nine Hox genes, but these derive from different classes than those found in *C. intestinalis*, indicating independent reductive evolution. *O. dioica* has *Hox1, 2, 4, 9* (two distinct copies), *10, 11, 12* and *13* orthologs and has thus lost *Hox3* independently and retained *Hox9* and *11*, compared to *C. intestinalis*.[30,31] Again, expression patterns of these retained Hox genes are organized along the anterior-posterior axis as expected, though that temporal expression patterns are disrupted. However in *O. dioica* the Hox cluster is no longer detectable. Cloning and sequencing of large-insert clones revealed that no *O. dioica* Hox gene was within ~200 kb of any other Hox gene and that the immediate genomic environment of each Hox gene included a wide range of different protein coding genes.[30] In this *O. dioica* resembles *C. elegans*, where Hox genes are dispersed across the genome.

From these two major exceptions to the rules of Hox cluster conservation, it is tempting to conclude that radical changes, especially simplifications, in body plans can permit Hox cluster breakdown and Hox gene loss.[30] This in turn suggests that perhaps the nematode ancestor was a more complex organism and that reconstructions of ecdysozoan and bilaterian ancestors with segmentation and appendages may not be too fanciful.

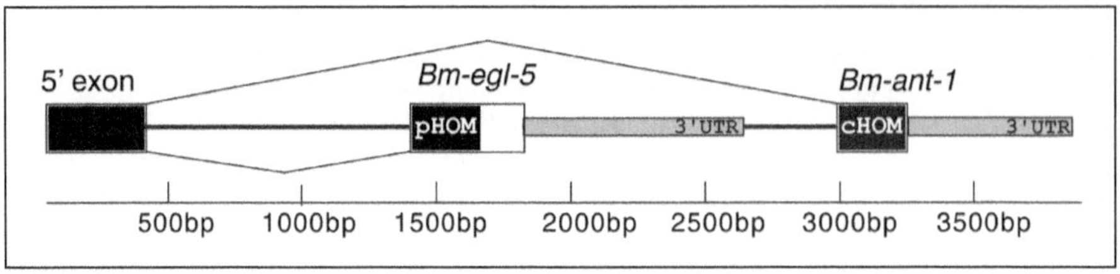

Figure 2. Alternative splicing of homeodomains in *Brugia malayi*. The *egl-5* homeodomain in *Brugia malayi* follows the same N-terminal domain as does a homeodomain belonging to an *Antennapedia* class HOX gene (*ant-1*). In the *B. malayi* genome this is achieved not by duplication of the 5′ exon encoding the N-terminal domain, but by alternative splicing of two homeodomain exons to the same 5′ exon. In *C. elegans* and relatives, only the *egl-5* homeodomain is present, suggesting that this arrangement may be an intermediate in the route to loss of the *Antennapedia*-like homeodomain. cHOM: central (*Antennapedia*) group homeodomain; pHOM: posterior group homeodomain; 3′UTR: 3′ untranslated region. Scale in bases. After Aboobaker and Blaxter (2003).

Hox Gene Loss in Flagrante

We also found an extra central class Hox6-8 gene in Clade I and Clade III species and once again the *T. spiralis* gene most closely resembled those of other phyla, while orthologs from Clade III nematodes were more divergent. Thus sometime after the divergence of Clade IV/V and Clade III groups this central class Hox homeodomain was lost.

Further investigation of this central class gene in *Brugia malayi* revealed that the homeodomain exon is spliced to the same 5′ exon as the homeodomain exon of the nematode posterior group gene *egl-5*. This 5′ exon in *B. malayi* contains regions of similarity to the 5′ exons of *egl-5* in *P. pacificus* and *C. elegans*, suggesting that it does derive from nematode *egl-5* genes. Thus, although the homeodomain of the central class gene has not been lost from Clade III nematodes, its N-terminal domain has (Fig. 2). We have since confirmed that this arrangement is also conserved in *A. suum* (and thus is not just a curiosity of just one Clade III species; Aboobaker et al, unpublished data). By performing semi-quantitative RT-PCR across the *B. malayi* life cycle we were also able to show that splicing of the single 5′ exon to the two different homeodomains was developmentally regulated. In particular, we observed that the *B. malayi egl-5* gene was expressed during the larval stages of development and not just embryonically.

This arrangement of alternatively spliced homeodomains linked to a shared N-terminal domain immediately suggests a molecular evolutionary mechanism for gradual loss of Hox gene coding sequence. It is possible that the divergent nature of the *egl-5* homeodomain is in part due to the fact it regulates targets that used to be regulated by multiple central class genes. This remains to be tested using functional studies in appropriate nematode species.

Nematode Hox Gene Function: A Story of Novelty, Conservation and Redeployment

In some respects nematodes have perhaps been underutilized as evolutionary developmental models. Their conserved simple body plans allow morphological homologies to be assigned almost entirely unambiguously. Most comparative genetic work has been with species within the same major clade as *C. elegans*,[32,33] so for example we know almost nothing about the functions of the extra Hox genes in other clades. This is due to the fact that as yet free-living models from other parts of the phylum await to be established.

Nonetheless we do know, in some cases in exquisite detail, the functions of the *C. elegans* Hox genes.[34-41] Taken together these suggest that they are still involved in specifying fate decisions, but within the context of a strict lineage-based developmental mode rather than specifying cells fates within spatial regions. *P. pacificus* has proven to be an exciting satellite model for understanding evolutionary developmental changes in some aspects of nematode development and the analyses of the functional changes involving the Hox genes *lin-39* and *mab-5* have been particularly elegant and informative.[11,12]

Hox gene function in *C. elegans* has some similarity to that in vertebrates and arthropods. Hox genes have roles in specifying cell fates along the body axis and to some extent follow the rules of spatial colinearity by ortholog group (even though their chromosomal order does not). They can also display cross- and auto-regulation.[42] The major difference is that expression in *C. elegans* is dependent on lineage rather than position, such that transplanted cells express Hox genes according to their lineage rather than their new position.[38,41]

It appears that of the six Hox genes in *C. elegans* only the anterior genes *ceh-13* and the posterior gene *php-3* are strictly required for embryogenesis.[43-45] Initial analyses of *nob-1* also suggested an early, essential role, but this was an artifact of the molecular lesions generated in *nob-1* analysed, which also deleted *php-3*.[45] The *ceh-13* gene is required for proper organisation of the anterior lateral epidermis and anterior body wall muscles. The identity of the affected cells remains correctly specified. *ceh-13* also has a role in male tail morphogenesis.[35] Elimination of *php-3* and *nob-1* from the *C. elegans* embryo results in gross posterior defects and the transformation of the fates of posterior lineages to those of more anterior lineages.[45] The remaining three Hox genes are not required for embryogenesis, with triple *lin-39*, *mab-5* and *egl-5* mutants surviving through embryogenesis, but are instead required for specification of cell fates and cell migrations in postembryonic development.[38]

Both *lin-39* and *mab-5* are used in defining the vulval equivalence group (VEG) of six ventral stem-cell neuroectoblasts from an initial field of twelve P cells. While *lin-39* is expressed in all the VEG stem cells, *mab-5* is only expressed in P8 and is part of the genetic machinery that restricts this P-cell's competence compared to the other VEG cells.[38] The specific fates of the equivalent *lin-39*-defined VEG cells are then set through interactions mediated by additional signals including EGF-Ras pathway signaling and Wnt signaling.[37,46] The vulva is formed from the posterior daughters of the P cells, while the anterior daughters produce neurons via specific division and differentiation patterns that are again controlled by *lin-39* and *mab-5* expression domains. In this post-embryonic developmental system, *C. elegans* Hox function is analogous to that observed for Hox genes in other taxa, in that it is part of a system for defining fates of fields of cells, but the cells involved are not colocalized in the embryo and indeed some neural daughters of the P cells migrate large distances from their sisters.[40] These migrations are under Hox control. In the male, in addition to patterning the fates of the neural daughters of the P cells, the Hox genes are involved in promoting particular cell fates in P cells participating in the hook equivalence group.

To function properly, *C. elegans* Hox genes require the nematode homologs of the Hox cofactors *extradenticle*/PBX (*ceh-20* and *ceh-40*) and *homothorax/MEIS* (*unc-62* and *psa-3*).[14,47-50] *C. elegans* Hox gene regulation by the Wnt and TGF-β signaling pathways may also represent conserved aspects of Hox gene function.[34,35,48,51] There is also evidence that *C. elegans* Hox gene regulation also involves chromatin remodeling complexes,[52] including members of the Polycomb group (PcG) proteins as is the case in both vertebrates and flies.[53-55] But in each of these cases there are novelties in the regulatory interactions with *C. elegans* Hox genes that suggest that much has changed, apart from just Hox gene loss. For example, the regulation of Hox genes by SOP-2 (which contains a SAM domain, also present in PcG proteins) suggests conservation, but in fact *sop-2* is not a true ortholog of any PcG genes in other animals.[54] Another example is the interaction of the MEIS ortholog *psa-3* with the Hox gene *nob-1*. There is no evidence that PSA-3 and NOB-1 interact directly: instead *nob-1* regulates the expression of *psa-3* through *nob-1/ceh-20* binding site present in a *psa-3* intron.[48] Thus not only have Hox genes been lost and undergone rapid sequence evolution, but the genetic networks in which they act have also experienced accelerated change. Indeed recent analyses of transcription factors in sequenced *Caenorhabditis* species genomes suggest rapid sequence evolution is a general feature.[56]

The evolution of function between *C. elegans* and *P. pacificus* Hox genes has been elegantly demonstrated. The *lin-39* gene is required for development of the nematode vulva in *C. elegans* and related nematodes. Detailed analysis of its function *P. pacificus* has revealed that significant evolutionary changes have occurred in how *lin-39* specifies the vulval cell lineage have occurred. In *P. pacificus*, *lin-39* defines the vulval equivalence group in a very similar way to its action in *C. elegans*, However, *P. pacificus* vulval precursor cells (VPCs) lacking *lin-39* activity undergo apoptosis (rather than cell fusion, as in *C. elegans*). If programmed cell death in *P. pacificus* is defective (such as in a *lin-39/ced-3* double mutant), a normal vulva is still formed.[9,11] Thus, unlike in *C. elegans*, *lin-39* is not required for vulval morphogenesis in *P. pacificus* and instead has a role in preventing apoptosis. Clearly this is a significant change.

Mutations in the other universal nematode central class gene *mab-5* are very similar in their effects on male phenotypes, with the homologous rays (R1-R6) failing to develop in both species. However the *P. pacificus mab-5* mutants have additional defects resulting in the induction of an ectopic vulva in a posterior position.[10,12] Clearly this role of *mab-5* evolved since the last common ancestor of the two species or has been lost in *C. elegans* while the role in male tail patterning has remained conserved.

Conclusion

Clearly the nematode case demonstrates an important point that we would do well not to neglect entirely in our consideration of the evolution of developmental processes. It is probably no coincidence that in other lineages in which Hox gene loss has occurred a loss of morphological and developmental complexity compared to related taxa (or phyla) is evident,[30] although we note that both Urochordata and Nematoda use lineage-driven developmental modes and thus the association may also be with this canalized developmental mechanism. While Hox gene functional evolution has clearly been important for morphological novelties in animals, Hox gene loss in nematodes is an extreme case of functional (d)evolution.

We would like to understand more about how these losses happened, about the functions of these genes before they were lost and whether losses were causal or a result of the evolution of the extant mode of development most evident in *C. elegans*. In order to this we would need to study the evolution and function of Hox genes in other nematodes with a more complete content of Hox genes. The genomic information of available for the Nematoda is increasing rapidly (Table 1) and we are steadily developing the capacity to apply functional genomic approaches to traditionally genetically inaccessible species. This will provide a rich resource for further study of gene evolution and function with respect to the unusual case of the nematode cluster.

Table 1. Genome sequence resources for Phylum Nematoda

Species	Order	Major Clade[1]	Genome Sequence
Caenorhabditis elegans	Rhabditida	V	Complete
Caenorhabditis briggsae	Rhabditida	V	Published first draft
Caenorhabditis remanei	Rhabditida	V	Online first draft[2]
Caenorhabditis japonica	Rhabditida	V	Online first draft[2]
Caenorhabditis brenneri	Rhabditida	V	Online first draft[2]
Haemonchus contortus	Strongylida	V	Online first draft[3]
Pristionchus pacificus	Diplogasterida	V	Published first draft
Meloidogyne hapla	Tylenchida	IV	Published first draft
Meloidogyne incognita	Tylenchida	IV	Published first draft
Brugia malayi	Spirurida	III	Published first draft
Trichinella spiralis	Trichocephalidae	I	Online first draft[2]

[1]Major clades are as defined by Blaxter et al (1998); [2]Available at http://genome.wustl.edu/; [3]Available at http://www.sanger.ac.uk/

Acknowledgements

We would like to thank Jean S. Deutsch for his encouragement and patience and colleagues in the nematode genome projects for early access to data.

References

1. Akam M. Hox genes: from master genes to micromanagers. Curr Biol 1998; 8(19):R676.
2. McGinnis W, Levine MS, Hafen E et al. A conserved DNA sequence in homoeotic genes of the Drosophila Antennapedia and bithorax complexes. Nature 1984; 308(5958):428.
3. Gehring WJ, Hiromi Y. Homeotic genes and the homeobox. Annu Rev Genet 1986; 20:147.
4. Aboobaker A, Blaxter M. Hox gene evolution in nematodes: novelty conserved. Curr Opin Genet Dev 2003; 13(6):593.
5. Burglin TR, Ruvkun G. The Caenorhabditis elegans homeobox gene cluster. Curr Opin Genet Dev 1993; 3(4):615.
6. Bürglin TR, Ruvkun G, Coulson A et al. Nematode homeobox cluster. Nature 1991; 351(6329):703.
7. The C. elegans Genome Sequencing Consortium, Genome sequence of the nematode C. elegans: a platform for investigating biology. Science 1998; 282(5396):2012.
8. Aboobaker AA, Blaxter ML. Hox gene loss during dynamic evolution of the nematode cluster. Curr Biol 2003; 13(1):37.
9. Sommer RJ, Eizinger A, Lee KZ et al. The Pristionchus HOX gene Ppa-lin-39 inhibits programmed cell death to specify the vulva equivalence group and is not required during vulval induction. Development 1998; 125(19):3865.
10. Jungblut B, Sommer RJ. The Pristionchus pacificus mab-5 gene is involved in the regulation of ventral epidermal cell fates. Curr Biol 1998; 8(13):775.
11. Grandien K, Sommer RJ. Functional comparison of the nematode Hox gene lin-39 in C. elegans and P. pacificus reveals evolutionary conservation of protein function despite divergence of primary sequences. Genes Dev 2001; 15(16):2161.
12. Gutierrez A, Knoch L, Witte H et al. Functional specificity of the nematode Hox gene mab-5. Development 2003; 130(5):983.
13. Streit A, Kohler R, Marty T et al. Conserved regulation of the Caenorhabditis elegans labial/Hox1 gene ceh-13. Dev Biol 2002; 242(2):96.
14. Van Auken K, Weaver D, Robertson B et al. Roles of the Homothorax/Meis/Prep homolog UNC-62 and the Exd/Pbx homologs CEH-20 and CEH-40 in C. elegans embryogenesis. Development 2002; 129(22):5255.
15. Sidow A, Thomas WK. A molecular evolutionary framework for eukaryotic model organisms. Current Biology 1994; 4:596.
16. Salser SJ, Kenyon C, Patterning C. elegans: homeotic cluster genes, cell fates and cell migrations. Trends in Genetics 1994; 10:159.
17. Wilson R, Ainscough R, Anderson K et al. 2.2 Mb of contiguous nucleotide sequence from chromosome III of C. elegans. Nature 1994; 368:32.
18. Mushegian AR, Garey JR, Martin J et al. Large-scale taxonomic profiling of eukaryotic model organisms: a comparison of orthologous proteins encoded by the human, fly, nematode and yeast genomes. Genome Res 1998; 8(6):590.
19. Aguinaldo AM, Turbeville JM, Linford LS et al. Evidence for a clade of nematodes, arthropods and other moulting animals. Nature 1997; 387(6632):489.
20. Halanych KM, Bacheller JD, Aguinaldo AM et al. Evidence from 18S ribosomal DNA that the lophophorates are protostome animals. Science 1995; 267(5204):1641.
21. Zheng J, Rogozin IB, Koonin EV et al. Support for the Coelomata clade of animals from a rigorous analysis of the pattern of intron conservation. Mol Biol Evol 2007; 24(11):2583.
22. Philip GK, Creevey CJ, McInerney JO. The opisthokonta and the ecdysozoa may not be clades: stronger support for the grouping of plant and animal than for animal and fungi and stronger support for the coelomata than ecdysozoa. Mol Biol Evol 2005; 22(5):1175.
23. Adoutte A, Balavoine G, Lartillot N et al. The new animal phylogeny: reliability and implications. Proc Natl Acad Sci USA 2000; 97(9):4453.
24. Dunn CW, Hejnol A, Matus DQ et al. Broad phylogenomic sampling improves resolution of the animal tree of life. Nature 2008; 452(7188):745.
25. Ruvkun G, Hobert O. The taxonomy of developmental control in Caenorhabditis elegans. Science 1998; 282(5396):2033.
26. de Rosa R, Grenier JK, Andreeva T et al. Hox genes in brachiopods and priapulids and protostome evolution. Nature 1999; 399(6738):772.
27. Blaxter ML, De Ley P, Garey JR et al. A molecular evolutionary framework for the phylum Nematoda. Nature 1998; 392(6671):71.

28. Meldal BH, Debenham NJ, De Ley P et al. An improved molecular phylogeny of the Nematoda with special emphasis on marine taxa. Mol Phylogenet Evol 2007; 42(3):622.
29. Ikuta T, Yoshida N, Satoh N et al. Ciona intestinalis Hox gene cluster: Its dispersed structure and residual colinear expression in development. Proc Natl Acad Sci USA 2004; 101(42):15118.
30. Seo HC, Edvardsen RB, Maeland AD et al. Hox cluster disintegration with persistent anteroposterior order of expression in Oikopleura dioica. Nature 2004; 431(7004):67.
31. Edvardsen RB, Seo HC, Jensen MF et al. Remodelling of the homeobox gene complement in the tunicate Oikopleura dioica. Curr Biol 2005; 15(1):R12.
32. Delattre M, Felix MA. Polymorphism and evolution of vulval precursor cell lineages within two nematode genera, Caenorhabditis and Oscheius. Curr Biol 2001; 11(9):631.
33. Kiontke K, Barrière A, Kolotuev I et al. Trends, stasis and drift in the evolution of nematode vulva development. Curr Biol 2007; 17(22):1925.
34. Teng Y, Girard L, Ferreira HB et al. Dissection of cis-regulatory elements in the C. elegans Hox gene egl-5 promoter. Dev Biol 2004; 276(2):476.
35. Stoyanov CN, Fleischmann M, Suzuki Y et al. Expression of the C. elegans labial orthologue ceh-13 during male tail morphogenesis. Dev Biol 2003; 259(1):137.
36. Shemer G, Podbilewicz B. LIN-39/Hox triggers cell division and represses EFF-1/fusogen-dependent vulval cell fusion. Genes Dev 2002; 16(24):3136.
37. Maloof JN, Kenyon C. The Hox gene lin-39 is required during C. elegans vulval induction to select the outcome of Ras signaling. Development 1998; 125(2):181.
38. Kenyon CJ, Austin J, Costa M et al. The dance of the Hox genes: patterning the anteroposterior body axis of Caenorhabditis elegans. Cold Spring Harb Symp Quant Biol 1997; 62:293.
39. Salser SJ, Kenyon C. A C. elegans Hox gene switches on, off, on and off again to regulate proliferation, differentiation and morphogenesis. Development 1996; 122(5):1651.
40. Harris J, Honigberg L, Robinson N et al. Neuronal cell migration in C. elegans: regulation of Hox gene expression and cell position. Development 1996; 122(10):3117.
41. Cowing D, kenyon C. Correct Hox gene expression established independently of position in Caenorhabditis elegans. Nature 1996; 382:353.
42. Ferreira HB, Zhang Y, Zhao C et al. Patterning of Caenorhabditis elegans posterior structures by the Abdominal-B homolog, egl-5. Dev Biol 1999; 207(1):215.
43. Brunschwig K, Wittmann C, Schnabel R et al. Anterior organization of the Caenorhabditis elegans embryo by the labial-like Hox gene ceh-13. Development 1999; 126(7):1537.
44. Wittmann C, Bossinger O, Goldstein B et al. The expression of the C. elegans labial-like Hox gene ceh-13 during early embryogenesis relies on cell fate and on anteroposterior cell polarity. Development 1997; 124(21):4193.
45. Van Auken K, Weaver DC, Edgar LG et al. Caenorhabditis elegans embryonic axial patterning requires two recently discovered posterior-group Hox genes. Proc Natl Acad Sci USA 2000; 97(9):4499.
46. Eisenmann DM, Maloof JN, Simske JS et al. The beta-catenin homolog BAR-1 and LET-60 Ras coordinately regulate the Hox gene lin-39 during Caenorhabditis elegans vulval development. Development 1998; 125(18):3667.
47. Takács-Vellai K, Vellai T, Chen EB et al. Transcriptional control of Notch signaling by a HOX and a PBX/EXD protein during vulval development in C. elegans. Dev Biol 2007; 302(2):661.
48. Arata Y, Kouike H, Zhang Y et al. Wnt signaling and a Hox protein cooperatively regulate psa-3/Meis to determine daughter cell fate after asymmetric cell division in C. elegans. Dev Cell 2006; 11(1):105.
49. Yang L, Sym M, Kenyon C. The roles of two C. elegans HOX cofactor orthologs in cell migration and vulva development. Development 2005; 132(6):1413.
50. Liu J, Fire A. Overlapping roles of two Hox genes and the exd ortholog ceh-20 in diversification of the C. elegans postembryonic mesoderm. Development 2000; 127(23):5179.
51. Maloof JN, Whangbo J, Harris JM et al. A Wnt signaling pathway controls hox gene expression and neuroblast migration in C. elegans. Development 1999; 126(1):37.
52. Guerry F, Marti CO, Zhang Y et al. The Mi-2 nucleosome-remodeling protein LET-418 is targeted via LIN-1/ETS to the promoter of lin-39/Hox during vulval development in C. elegans. Dev Biol 2007; 306(2):469.
53. Zhang H, Smolen GA, Palmer R et al. SUMO modification is required for in vivo Hox gene regulation by the Caenorhabditis elegans Polycomb group protein SOP-2. Nat Genet 2004; 36(5):507.
54. Zhang H, Azevedo RB, Lints R et al. Global Regulation of Hox Gene Expression in C. elegans by a SAM Domain Protein. Dev Cell 2003; 4(6):903.
55. Ross JM, Zarkower D. Polycomb Group Regulation of Hox Gene Expression in C. elegans. Dev Cell 2003; 4(6):891.
56. Haerty W, Artieri C, Khezri N et al. Comparative analysis of function and interaction of transcription factors in nematodes: extensive conservation of orthology coupled to rapid sequence evolution. BMC Genomics 2008; 9:399.

Chapter 8

Are the Deuterostome Posterior Hox Genes a Fast-Evolving Class?

Robert Lanfear*

Abstract

There has been a great deal of interest in analysing the molecular evolution of the Hox cluster using both bioinformatic and experimental approaches. The posterior Hox genes have been of particular interest to both groups of biologists for a number of reasons: they appear to be associated with the evolution of a number of morphological novelties; the protostomes appear to be have lost a highly-conserved and functionally important amino acid motif (the hexapeptide motif) from their posterior Hox genes; and deuterostome posterior Hox genes seem to be evolving more quickly than all other Hox genes. In this chapter I will discuss the last of these points.

The idea that Deuterostome posterior Hox genes were evolving more quickly than other Hox genes was first suggested by David Ferrier and colleagues.[1] In this chapter, I start by introducing the posterior Hox genes—their distribution among the animal phyla and the likely sequence of duplications that led to this distribution. I then introduce the idea of 'deuterostome posterior flexibility'[1] and examine this hypothesis in light of more recent phylogenetic and genomic work on the Hox cluster. Finally, I discuss some new approaches that could be used to test directly for differential rates of evolution among Hox genes and to assess what might underlie these differences.

The Distribution of the Posterior Hox Genes in the Metazoa

The posterior Hox genes exist in all the major bilaterian phyla examined so far, as well as in the Cnidaria (Fig. 1). To date no Hox genes of any kind have been found in any other phyla (either metazoan or otherwise), thus it seems reasonable to assume that the posterior Hox genes came into existence after the divergence of the poriferan lineage, but before the divergence of the Cnidaria and the other Metazoan phyla, roughly 650-850 million years ago.[2] Broadly speaking, the posterior Hox genes of the bilaterian phyla can be resolved into three major groupings, which are delineated along the same lines as the 'new' animal phylogeny[3,4] (Fig. 1): the Deuterostomia (chordates, echinoderms etc.) possess orthologs of *Hox9* to *Hox15* genes; and within the Protostomia the Lophotrochozoa (annelids, molluscs etc.) possess orthologs of the *Post-1* and *Post-2* genes; and the Ecdysozoa (insects, nematodes etc.) possess orthologs of the *Abd-B* gene. The posterior Hox genes of the acoel flatworms and the Cnidaria do not group robustly with any of the major groupings described above, although it is well established that they are indeed posterior Hox genes.[5-10] Despite occasional difficulties in assigning Hox genes to one of these three groupings, the major bilaterian groupings of posterior Hox genes have been repeatedly confirmed by different phylogenetic studies,[1,11-15] and are considered so robust that the possession of one type of posterior Hox gene or another is now considered good evidence on which to base the phylogenetic affinity of otherwise enigmatic taxa.[4,13] Unfortunately,

*Robert Lanfear—Centre for Macroevolution and Macroecology, School of Botany and Zoology, Building 116 Daley Road, Australian National University, ACT 0200, Australia.
Email: rob.lanfear@anu.edu.au

Hox Genes: Studies from the 20th to the 21st Century, edited by Jean S. Deutsch.

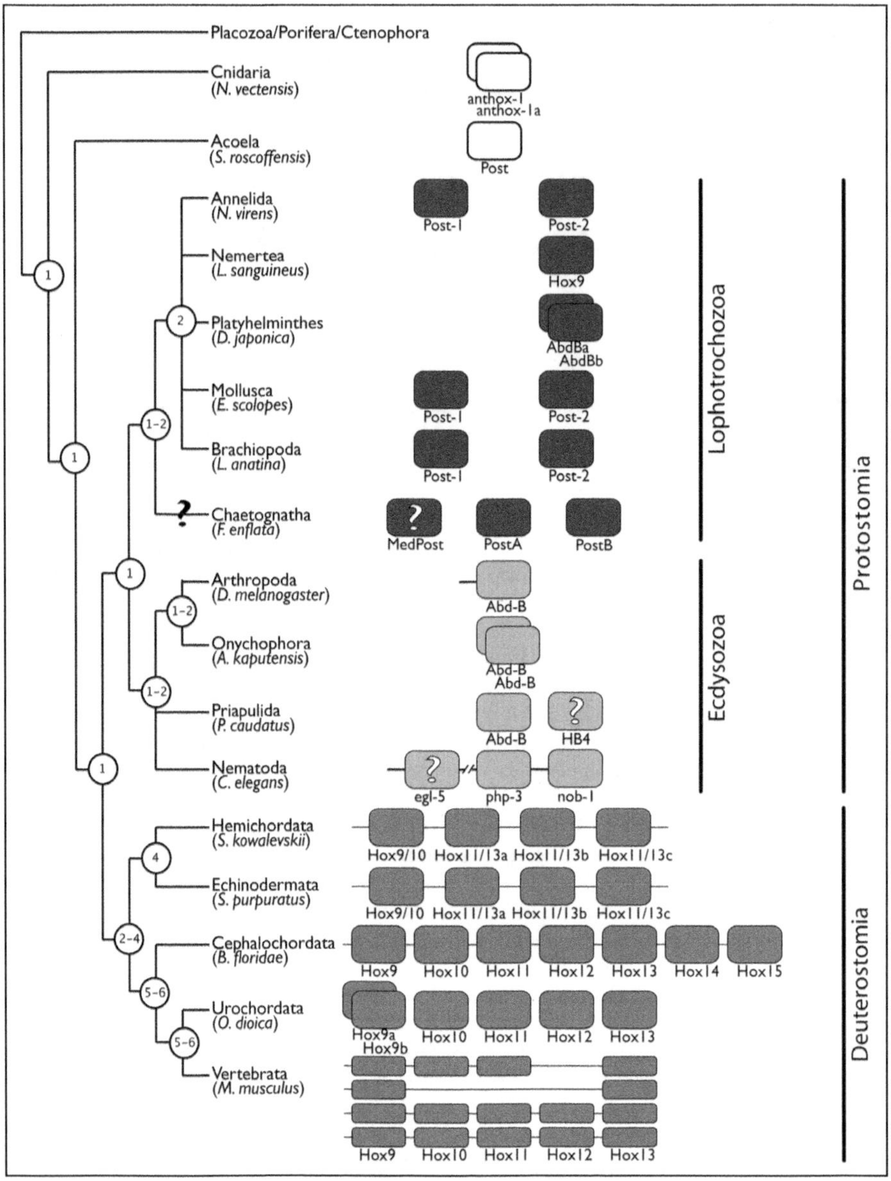

Figure 1. A sketch of the evolutionary history and current distribution of posterior Hox genes. Shading indicates approximate orthology relationships. Overlaid boxes (e.g., *Hox9a* and *Hox9b* in Urochordates) indicate recent duplication events. Question marks in boxes represent uncertain orthology relationships (see text) and question marks on the phylogenetic tree represent uncertain phylogenetic relationships. Where linkage relationships are known, they are indicated by connecting lines between boxes. Data references are as follows: Cnidaria,[6] Acoela,[5] Annelida,[12] Nemertea,[16] Platyhelminthes,[17] Mollusca,[18] Brachiopoda,[12] Chaetognatha,[13] Arthropoda,[19] Onychophora,[20] Priapulida,[12] Nematoda,[21] Hemichordata,[15] Echinodermata,[11] Cephalochordata,[22] Urochordata,[23] Vertebrata.[24]

the more or less robust grouping of many different types of posterior Hox genes is not reflected in their nomenclature and more often than not the existence of two posterior Hox genes with the same name is no indication of their relatedness (see e.g., Fig. 1 in which the major orthology groups are indicated by shading and the names are listed underneath).

Problematic Assignments of Hox Genes as 'Posterior'

Despite the relatively simple sketch of the distribution of posterior Hox genes given above, there are a number of instances in which the classification of a Hox gene as 'posterior' remains uncertain (indicated in Fig. 1 with a '?'). In some cases, orthology assignment is problematic because only very short fragments of the Homeobox have been sequenced and isolated.[25,26] In other cases however the situation can be somewhat more complex and the analysis of the whole homeodomain as well as its flanking sequences has proved insufficient to confidently ascertain whether some genes are posterior Hox genes at all, let alone to decide whether they fall into any of the three major groupings of posterior Hox genes described above. A case in point is the posterior Hox genes of the cnidarian *Nematostella vectensis*. Initial phylogenetic analyses of homeodomain sequences suggested that this species did not contain any true posterior Hox genes at all,[7] however a recent (and more thorough) re-analysis of precisely the same dataset, using the same phylogenetic procedures, suggests the opposite.[6] Despite disagreements about specific cases, however, it is well accepted that there exist a number of other posterior Hox genes in the Cnidaria.[10]

Another problematic case is the *MedPost* genes of chaetognaths. The homeodomains of these genes contain diagnostic residues of both the median (namely Q6, T7 and E59 and the LTR(R/K) RRI peptide at positions 26-32) and posterior (K3, A14, R18, Y20, Q36) Hox genes.[26] They were thus initially suggested to be mosaic genes that had arisen prior to the divergence of true posterior Hox genes from the other Hox genes. On the basis of this and the failure to find any unambiguous posterior Hox genes in chaetognaths, it was suggested that chaetognaths may have diverged from the bilaterians before the protostome/deuterostome split.[26] Recently however, a *Medpost* ortholog and two true Posterior genes (*PostA* and *PostB*) have been discovered in a different chaetognath species.[13] The discovery of true Posterior Hox genes (although they are difficult to resolve to one of the three major classes of posterior Hox genes mentioned above) suggests that the *MedPost* genes are likely to be a chaetognath-specific innovation, although their origin remains obscure. It is possible that they will end up being classified as true posterior Hox genes on the basis of data other than the sequence alone (e.g., data on their position in the cluster and on their developmental role).

Finally, the nematode Hox gene *egl-5* has also been the subject of some controversy. Although a number of studies have suggested that *egl-5* is a posterior Hox gene based on sequence analysis and its position in the remnants of the *C. elegans* Hox cluster,[21,27-29] others have suggested that *egl-5* cannot be classified as a posterior Hox gene with any certainty.[4,12,30]

Early Duplications of the Posterior Hox Genes

In order to examine whether the deuterostome posterior Hox genes are a fast-evolving class, it is helpful to first clarify the sequence of duplications that led to the current distribution of posterior Hox genes in the extant taxa. Figure 1 shows an attempt to do this, with predicted numbers of posterior Hox genes marked onto ancestral nodes of the tree. Three types of uncertainty limit the accuracy of this procedure: uncertainty in the phylogenetic placement of certain taxa (e.g., the chaetognaths), uncertainty as to the relationships between different posterior Hox genes (e.g., the *Hox9-15* genes of cephalochordates and the *Hox9-14* genes of most vertebrates) and uncertainty as to the classification of some genes as posterior Hox genes (see above).

There is another problem inherent in the estimation of ancestral gene content, which is distinct from those listed above—there is very likely to be an ascertainment bias in our knowledge of the distribution of posterior Hox genes among different taxonomic groups. The majority of Hox genes have been discovered by PCR surveys or the screening of genomic libraries, both of which are limited techniques insofar as they are only able to recover sequences that are sufficiently similar to other known sequences. Because of this, it has often been the case that initial PCR surveys greatly underestimate the number of Hox genes in a given species. As such, we can only be sure of the Hox gene content of a given species once a fully assembled genome sequence is available and even when this is the case, current problems with whole-genome assembly methods mean that it is preferable to double-check the assembly using genomic walking. These methodological issues are neatly illustrated by the recent discovery of the amphioxus *Hox15* gene. This gene had gone undiscovered until the recent completion of the amphioxus genome, despite the fact that the amphioxus Hox cluster is among the most thoroughly studied of all Hox clusters[1,22,31] and that a previous study which had explicitly set out to look for a *Hox15* gene in amphioxus had concluded that it didn't exist.[32] This ascertainment bias in Hox gene identification will tend to favour the discovery of Hox genes in those clades for which we have more genome sequences—both due to the direct identification of Hox genes from the genome sequences themselves and by the indirect use of those genome sequences to fine-tune methods of 'fishing' for Hox genes in closely related species. Therefore, it is possible that a proportion of the excess of posterior Hox genes known in deuterostomes might be due to the effects of ascertainment bias in this clade.

Despite the difficulties inherent in such a procedure, it is still possible to estimate the posterior Hox gene complement of hypothetical ancestral species at important points in the history of the Metazoa. The number of posterior Hox genes at each ancestral node in Figure 1 was estimated by comparing phylogenetic trees of Hox genes from various sources and from the discussions of previous authors.[5,6,8,10-12,14,15,18,20,22,30,33-41] For instance the chordate ancestor (Fig. 1), likely posessed 5 or 6 genes (although other numbers are also conceivable): a minimum of 5 posterior Hox genes seems probable since all chordates have at least 5 posterior Hox genes which tend to group together (although with little resolution) on phylogenetic trees. However there is some evidence that the chordate ancestor may have possessed 6 posterior Hox genes as both cephalochordates and some vertebrates[42,43] possess a *Hox14* and it is quite possible (though difficult to show with any degree of certainty) that the vertebrate and cephalochordate *Hox14* genes are homologous and that the chordate ancestor therefore also possessed a *Hox14*.[37]

The 'Deuterostome Posterior Flexibility' Hypothesis

"Deuterostome posterior flexibility" refers to the hypothesis that the posterior Hox genes of Deuterostomes are evolving at a faster rate than other Hox genes.[1] This hypothesis was put forward to explain the fact that in phylogenetic analyses the posterior Hox genes of deuterostomes (*Hox9+*) tend to be poorly resolved, whereas the posterior Hox genes of protostomes tend to resolve with high support (into the *AbdB*-like genes for the Ecdysozoa and the *Post1*-like and *Post2*-like genes for the Lophotrochozoa). In this section, I introduce the original observations that the deuterostome posterior flexibility hypothesis was put forward to explain and discuss this hypothesis in light of recent phylogenetic and genomic studies.

Ferrier et al[1] undertook a genomic walk along the Amphioxus Hox cluster and discovered four new posterior Hox genes—*AmphiHox11*, *AmphiHox12*, *AmphiHox13* and *AmphiHox14*. Phylogenetic analysis of a large dataset of posterior Hox genes was carried out using maximum parsimony (MP) and neighbour-joining (NJ). These analyses showed that groupings of *AbdB*-like genes from the Ecdysozoa and *Post1*-like and *Post2*-like genes from the Lophotrochozoa were recovered with high bootstrap support. In contrast there was very low support for the grouping together of the deuterostome posterior Hox genes—individual orthology groups from within the vertebrates (e.g., vertebrate *Hox12*) were recovered with high support, but support for clustering of these groups with any other deuterostome posterior Hox genes was almost always less than 50%, well below the levels usually required for confident phylogenetic inference. The authors explored two possible evolutionary hypotheses for the origin of the amphioxus posterior Hox genes using a maximum likelihood (ML) based statistical significance test. The first hypothesis was that the amphioxus posterior Hox genes had arisen independently after the split of the amphioxus and vertebrate lineages and the second was that each amphioxus gene was orthologous to a vertebrate gene (e.g., *AmphiHox10* is orthologous to vertebrate *Hox10*). Interestingly, the first hypothesis (independent duplication) was significantly rejected, whereas the second hypothesis (orthologous genes) was statistically indistinguishable from the ML tree.[1] Thus, the most parsimonious explanation by far (with respect to the number of gene duplication and loss events that have to be postulated to explain a given phylogenetic tree) is that the chordate ancestor possessed copies of *Hox9, Hox10, Hox11, Hox12* and *Hox13* and that amphioxus and the vertebrates each inherited copies of these genes. The puzzle therefore was why the protostome groupings of posterior Hox genes (e.g., the *AbdB*-like genes) could be recovered with high confidence, whereas the deuterostome posterior Hox genes could not. Ferrier and colleagues resolved this dilemma by suggesting that deuterostome posterior Hox genes were evolving at a faster rate than other Hox genes. A faster rate of evolution would in turn have led to a faster degradation of phylogenetic signal in these genes and could therefore explain their lack of resolution in phylogenetic trees.

Recent Analyses Broadly Support the Posterior Flexibility Hypothesis

The enormous interest in the evolution of the Hox cluster has meant that a number of recent studies have performed comparable phylogenetic analyses to those in the Ferrier et al[1] study in which the deuterostome posterior flexibility hypothesis was proposed. Five studies in particular have included a wide representation of metazoan posterior Hox genes and reported measures of clade support such as nonparametric bootstrap proportions (BP) or Bayesian posterior probabilities (BPP).[11,14,15,22,23] Three of these studies include a dataset sufficient to compare the phylogenetic resolution of the posterior Hox genes of protostomes (i.e., the *AbdB*-like, *Post1*-like and *Post2*-like genes) to the phylogenetic resolution of the posterior Hox genes of deuterostomes (i.e., the *Hox9+* genes). All of these three studies support the observation that the resolution of the protostome posterior Hox genes is far higher than that of the deuterostome posterior Hox genes.[11,14,15] Additionally, two more studies support the notion that there is low resolution among the deuterostome posterior Hox genes relative to the deuterostome anterior Hox genes, although neither of these include sufficient data to compare this to the resolution of the protostome Hox genes.[22,23] These five studies do not represent five completely independent tests of the deuterostome posterior flexibility hypothesis as all of the studies use somewhat similar datasets and methods. However, it has recently been shown that conclusions drawn from phylogenetic analyses of homeodomains can be extremely sensitive to small changes in dataset composition and phylogenetic methodology,[44] so the agreement of all comparable studies to date lends credence to some important aspects of the hypothesis.

Despite the broad support for the deuterostome posterior flexibility hypothesis in recent studies, one interesting caveat to the hypothesis has emerged—that the hypothesis might not apply to all posterior Hox genes in all deuterostome taxa.[14,15] Both echinoderms and hemichordates have at least four posterior Hox genes (*Hox9/10, Hox11/13a, Hox11/13b* and *Hox11/13c*; see Fig. 1) whose similar nomenclature in the two taxa represents the likelihood that they were all inherited from a common ancestor (although this is disputed[22]). Two of these genes (*Hox9/10* and *Hox11/13a*) seem to show phylogenetic resolution consistent with the posterior flexibility hypotheses—i.e., they are poorly resolved. The other two genes (*Hox11/13b* and *Hox11/13c*) however, have been shown to group together with strong support in recent analyses: Holland et al[22] report a BPP of 0.92, a ML BP of 95% and a NJ BP of 99% for the grouping of *Hox11/13b* and *Hox11/13c* sequences and Cameron et al[11] report a NJ BP of 88% for the same grouping. Interpretation of this situation is complicated by the lack of resolution within the *Hox11/13b* and *Hox11/13c* clade. It might be the case that an ancestral *Hox11/13b* gene duplicated independently and recently in the hemichordate and echinoderm lineages to form the *Hox11/13c* genes. This explanation is consistent with the deuterostome posterior flexibility hypothesis, in that the high support for the grouping can reconciled with fast rates of molecular evolution by the postulation of a recent duplication. Another explanation for the same pattern, preferred by some authors,[14,15] is that both *Hox11/13b* and *Hox11/13c* were present in the hemichordate/echinoderm ancestor. This hypothesis is not consistent with deuterostome posterior flexibility, in that it requires these two genes to have been evolving much more slowly than other deuterostome posterior Hox genes (and thus retaining a greater proportion of their phylogenetic signal than other deuterostome posterior Hox genes, since both the hemichordate/echinoderm and the cephalochordate/vertebrate splits are predicted to have occurred at around the time of the Cambrian explosion[45,46]). Distinguishing among these possibilities will require detailed statistical tests of phylogenetic topologies, in order to compare trees consistent with each hypothesis.

In addition to phylogenetic studies, two recent genomic studies[47,48] have made observations that are consistent with the deuterostome posterior flexibility hypothesis. Both of these studies have shown that the intergenic regions between the posterior Hox genes tend to be less conserved than those between the anterior Hox genes. This is consistent with the notion that not only the coding sequences but also the regulatory regions of deuterostome posterior Hox genes are evolving at a faster rate than those of the deuterostome anterior and central Hox genes.

The Mechanistic Basis of Deuterostome Posterior Flexibility

In essence the deuterostome posterior flexibility hypothesis is based upon the observation that most deuterostome posterior Hox genes appear less well resolved than other Hox genes in molecular phylogenetic analyses. Explanations for this pattern of phylogenetic support can be split into two broad categories. The first category of explanations presupposes that most deuterostome posterior Hox genes are evolving faster than most other Hox genes and goes on to propose possible reasons why this might be the case. The second category of explanations proprose reasons why the observed patterns of phylogenetic support might have arisen in the absence of differential rates of evolution. Each of these categories is discussed in more detail below.

Faster Rates May Be Linked to Gene Duplications

There are a number of mechanisms that have been suggested to underlie a faster rate of molecular evolution in the deuterostome posterior Hox genes, of which perhaps the most convincing is that the increased rate is linked to gene duplication events.[1] Although exact numbers are hard to estimate (see Fig. 1), it is clear that there have been significantly more duplications of posterior Hox genes in the deuterostome lineage than in the protostome lineage. Following a gene duplication event, the most likely outcome is that one of the two 'daughter' genes quickly degenerates to become a pseudogene through the acquisition of deleterious mutations ('nonfunctionalisation'). It is also conceivable (though unlikely) that one of the two daughter genes acquires a beneficial mutation that confers a new function ('neo-functionalisation'). A third possibility is that the two daughter genes evolve in such a way that the functional repertoire of the original gene is divided between them ('sub-functionalisation', also known as the duplication-degeneration-complementation model).[49-52] Gene duplications can lead to an increase in the rate of molecular evolution in two ways. First, there may be a brief period of relaxed selective constraint immediately following a duplication event.[51,53] Second, both neo- and sub-functionalisation—which are likely to have occurred in the majority of the posterior Hox genes present in the extant taxa, by virtue of the fact that the genes are still operative—imply a period of positive selection as the genes evolve to operate with a new or subdivided functional repertoire.[54] Indeed, although it might well be impossible to demonstrate whether there had been relaxed or positive selection following ancient Hox gene duplications[55] (such as those duplications which created many of the posterior Hox genes), there is good evidence that positive selection has occured after more recent Hox-gene duplications.[49,56-58] Thus, given the excess of gene duplications in the posterior Hox genes of deuterostomes relative to other taxonomic groups, it is likely that there exists a link between these duplications and an increased rate of molecular evolution, thus potentially explaining the observation of low phylogenetic resolution among deuterostome posterior Hox gene sequences.

Faster Rates May Be Linked to Morphological Evolution

Another prominent feature in the recent literature is the hypothesised link between the evolution of Hox genes and the evolution of morphological novelties.[59-63] With respect to the deuterostome posterior flexibility hypothesis, there are tantalising correlations between the expansion of the chordate Hox cluster and the evolution of a chordate-specific features.[59] Among other novelties, the posterior Hox genes are involved in the patterning of the post-anal tail of all chordates,[64] the limbs and digits of vertebrates[65-69] and have been implicated in the evolution of the pelvis—a key adaptation for the tetrapod lineage.[64] Morphological novelties, and the genes that are involved in patterning them, might have fast rates of evolution for two reasons. First, both genes and morphology might be evolving under strong positive selection. Second the novel morphological features may not be as tightly constrained—either in a developmental or an evolutionary sense—as many other morphological features and thus many more mutations which affect the patterning genes are likely to be selectively neutral. In contrast to the posterior Hox genes, it has been argued that the structures that the anterior and central Hox genes are responsible for patterning tend to be highly constrained (e.g., the neural tube of cephalochordates and the rhombocephalon of vertebrates).[47,48]

It will be very difficult to test effectively whether there exists (or existed) a causal link between the rate of evolution of the posterior Hox genes and the development of new morphological features in certain taxa. However, there are two lines of evidence that are suggestive of such a link. First, the observation that the intergenic regions of deuterostome posterior Hox genes tend to be less conserved than the intergenic regions of other deuterostome Hox genes provides good evidence that the posterior Hox genes of chordates are less tightly constrained than either the anterior or central Hox genes. This observation seems to square well with the degree of evolutionary constraint of the structures which these genes pattern.[47,48] Second, the echinoderm/hemichordate clade is thought to have inherited a small post-anal extension from the deuterostome ancestor,[70] a feature that has been lost or obscured in echinoderms[71] and remains un-elaborated in hemichordates (despite the expression of all three hemichordate posterior Hox genes in this region[70]). It is interesting that the posterior Hox genes of these phyla are those that appear to buck the deuterostome posterior flexibility trend and have a slower rate of evolution than other deuterostome posterior Hox genes (see above). At present, this is just a coincidental observation, but genomic studies of the degree of conservation of intergenic regions in the hemichordate/echinoderm clade would be extremely informative with respect to a possible link between the rates of molecular and morphological evolution.

Processes Other Than Faster Rates Might Be Operating

It is possible that the observations that led to the proposal of the deuterostome posterior flexibility hypothesis could be explained without the need to posit differential rates of molecular evolution among Hox genes. No studies have explicitly compared the rates of evolution of different Hox genes. Indeed, in those cases where molecular branch lengths have been included in published analyses of Hox genes, there is no obvious trend for the deuterostome posterior Hox genes to have significantly longer branches than other Hox genes[22,23] as would be expected if they were evolving at a faster rate. It has been suggested that the observed patterns of phylogenetic support might be the result of nonphylogenetic signal in the data, rather than the result of differential rates of evolution among Hox genes.

Some models of the evolution of the Hox genes have been suggested in which there were long periods of stasis in the evolutionary history of certain genes.[26,30] Such periods of stasis contravene the assumption that phylogenetic distance will tend to increase with time. Although the implications of this for phylogenetic analyses have not been worked out in detail, it is conceivable (though perhaps unlikely) that such periods of stasis could contribute to the observed patterns of phylogenetic support among Hox genes.

A more plausible source of nonphylogenetic signal that could confound phylogenetic analyses involves the co-evolution of interacting proteins. It has been suggested that in those cases where a group of genes interact with a given protein (for instance vertebrate posterior Hox genes all interact with Meis1 proteins[72]), changes in the given protein (Meis1 in this case) within a given lineage might lead to correlated changes in all of the interacting proteins (the posterior Hox genes in this case) in that lineage.[73] This is problematic for conventional phylogenetic analyses as a fundamental assumption of such approaches is that all genes are evolving independently in all lineages. Simulations suggest that in those cases where the gene duplications are ancient and the evolutionary rate of the given protein is slow relative to the interacting proteins (as might be the case for the posterior Hox genes and Meis1 respectively) a conventional phylogenetic analyses of the duplicated genes will tend to be poorly resolved.[73] Thus it is feasible, although it remains untested, that this kind of process might explain the observed pattern of phylogenetic resolution among the Hox genes.

Conclusion and Future Directions

In the eight years since it was proposed that the deuterostome posterior Hox genes might be a fast evolving class a great many new Hox gene sequences have been published and a number of genomic studies of Hox genes have been undertaken. Concomitantly, our understanding of phylogenetic methodology, genomics and molecular evolution has increased significantly. However, despite these advances it is still difficult to come up with a reliable answer to the question: "Are the deuterostome posterior Hox genes a fast evolving class?" In general the available evidence weighs in favour of the idea that the majority of deuterostome posterior Hox genes are fast-evolving and the most likely mechanistic explanation for this is (in my opinion) that it is largely a result of the effects of gene duplication. It is difficult to make more concrete conclusions than this as there are a number of key deficiencies in the available data which preclude taxonomically broad-scale comparisons of the rates of evolution of different Hox genes and thus also preclude meaningful comparisons of the mechanistic underpinnings of such rate variation. Below I indicate where the current deficiencies in our understanding lie and suggest some approaches that might be taken to remove these deficiencies.

To date there have been no studies which have explicitly measured the rates of evolution of different Hox genes in different metazoan lineages. A comparative study of substitution rates in Hox genes is particularly important since it has been argued that not all deuterostome posterior Hox genes are fast evolving and that it is instead a phenomenon limited to the chordates.[14,15] The data and the methods to conduct a comparative study of rates of molecular evolution in the Hox genes are already available, although their application will be complicated by the very short alignable (60 amino acid) regions of different Hox genes.[74] Nevertheless, it might be possible to circumvent these difficulties by estimating the absolute rates of evolution of posterior Hox genes of closely related taxa using a dated molecular phylogeny and then comparing these absolute rates between different genes and taxa.

If a method can be found which allows the rates of evolution of different Hox genes in different lineages to be measured reliably, it may also be possible to compare the extent to which different putative explanatory variables (e.g., morphological evolution or gene duplication events) might be responsible for the observed variation in rates. For instance, methods which have been developed to test for links between rates of molecular evolution and speciation rates[75,76] could be adapted to test for a link between rates of molecular evolution and gene duplication events. Currently available methods to test for a link between morphological and molecular rates of evolution[77,78] would be much harder to apply to the Posterior Hox genes, however if such a study were carried out it would be the first study of its kind to systematically compare the rates of molecular evolution of developmental genes with the rates of evolution of the morphological features that those genes are responsible for patterning.

It is always difficult to rule out systematic bias in phylogenetic studies and it has been suggested that this might be a particular problem for studies of the Hox genes.[73] In particular, it has been suggested that co-evolutionary dynamics among Hox genes may confound conventional phylogenetic analyses, but thankfully there are existing methods that could be used to test for the existence of such processes in Hox genes.[73,79] If robust conclusions are to be made about differential rates of evolution in Hox genes in the absence of comparative studies of substitution rates, it will be important to carry out such tests.

The results of genomic studies indicate that the intergenic regions of the chordate posterior Hox genes are evolving more quickly than those of the other Hox genes of chordates.[47,48,80,81] Although this observation is certainly consistent with the idea that the deuterostome posterior Hox genes are a fast evolving class, it is insufficient to assess whether the posterior Hox genes of all deuterostomes are evolving more quickly than all of the other deuterostome Hox genes and these data are also uninformative with respect to the relative rates of deuterostome Hox genes to Hox genes from other phyla. An extension of the genomic approach to the rest of the deuterostomes (i.e., the hemichordates and echinoderms) and to nondeuterostome taxa will be important in this respect. It would already be possible to carry out comparable studies on a number of publically available protostome genomes, although given the current limitations of the methodology[31] it might prove to be the case that some of the currently available genome sequences are too divergent, or the Hox clusters too large, for such methods to be applicable.

Finally, current analyses of the evolution of Hox clusters can be somewhat hampered by the difficulty of assigning Hox genes to particular orthology groups. This is a particular problem with the deuterostome posterior Hox genes and a key area of work in this respect is further sequencing of the Hox clusters of key deuterostome taxa—in particular the lamprey and hagfish and the *Xenoturbella*—which it might be hoped will further elucidate the evolutionary history of the Hox clusters of deuterostomes.

References

1. Ferrier DEK, Minguillon C, Holland PWH et al. The amphioxus hox cluster: deuterostome posterior flexibility and hox14. Evol Dev 2000; 2(5):284-293.
2. Peterson KJ, Butterfield NJ. Origin of the eumetazoa: testing ecological predictions of molecular clocks against the proterozoic fossil record. Proc Natl Acad Sci USA 2005; 102(27):9547-9552.
3. Aguinaldo AM, Turbeville JM, Linford LS et al. Evidence for a clade of nematodes, arthropods and other moulting animals. Nature 1997; 387(6632):489-493.
4. Balavoine G, de Rosa R, Adoutte A. Hox clusters and bilaterian phylogeny. Mol Phylogenet Evol 2002; 24(3):366-373.
5. Cook CE, Jimenez E, Akam M et al. The hox gene complement of acoel flatworms, a basal bilaterian clade. Evol Dev 2004; 6(3):154-163.
6. Ryan JF, Mazza ME, Pang K et al. Pre-bilaterian origins of the hox cluster and the hox code: evidence from the sea anemone, nematostella vectensis. PLoS One 2007; 2:e153.
7. Kamm K, Schierwater B, Jakob W et al. Axial patterning and diversification in the cnidaria predate the hox system. Curr Biol 2006; 16(9):920-926.
8. Finnerty JR, Martindale MQ. Ancient origins of axial patterning genes: Hox genes and ParaHox genes in the cnidaria. Evol Dev 1999; 1(1):16-23.
9. Finnerty JR, Martindale MQ. Homeoboxes in sea anemones (cnidaria; anthozoa): A PCR-based survey of nematostella vectensis and metridium senile. Biol Bull 1997; 193(1):62-76.
10. Ferrier DEK, Holland PWH. Ancient origin of the hox gene cluster. Nat Rev Gen 2001; 2(1):33-38.
11. Cameron RA, Rowen L, Nesbitt R et al. Unusual gene order and organization of the sea urchin hox cluster. J Exp Zoolog B Mol Dev Evol 2006; 306(1):45-58.
12. de Rosa R, Grenier JK, Andreeva T et al. Hox genes in brachiopods and priapulids and protostome evolution. Nature 1999; 399(6738):772-776.
13. Matus D, Halanych KM, Martindale MQ. The hox gene complement of a pelagic chaetognath, flaccisagitta enflata. Integ Comp Biol 2007; 47:854-864.
14. Peterson KJ. Isolation of hox and parahox genes in the hemichordate ptychodera flava and the evolution of deuterostome hox genes. Mol Phylogenet Evol 2004; 31(3):1208-1215.
15. Aronowicz J, Lowe CJ. Hox gene expression in the hemichordate saccoglossus kowalevskii and the evolution of deuterostome nervous systems. Integ Comp Biol 2006; 46:890-901.
16. Kmita-Cunisse M, Loosli F, Bierne J et al. Homeobox genes in the ribbonworm lineus sanguineus: Evolutionary implications. Proc Natl Acad Sci USA 1998; 95(6):3030-3035.
17. Nogi T, Watanabe K. Position-specific and noncolinear expression of the planarian posterior (abdominal-B-like) gene. Dev Growth Differ 2001; 43(2):177-184.

18. Callaerts P, Lee PN, Hartmann B et al. HOX genes in the sepiolid squid euprymna scolopes: implications for the evolution of complex body plans. Proc Natl Acad Sci USA 2002; 99(4):2088-2093.
19. Akam M. Hox and HOM: homologous gene clusters in insects and vertebrates. Cell 1989; 57(3):347-349.
20. Grenier JK, Garber TL, Warren R et al. Evolution of the entire arthropod hox gene set predated the origin and radiation of the onychophoran/arthropod clade. Curr Biol 1997; 7(8):547-553.
21. Van Auken K, Weaver DC, Edgar LG et al. Caenorhabditis elegans embryonic axial patterning requires two recently discovered posterior-group hox genes. Proc Natl Acad Sci USA 2000; 97(9):4499-4503.
22. Holland LZ, Albalat R, Azumi K et al. The amphioxus genome illuminates vertebrate origins and cephalochordate biology. Genome Res 2008; 18(7):1100-1111.
23. Seo HC, Edvardsen RB, Maeland AD et al. Hox cluster disintegration with persistent anteroposterior order of expression in oikopleura dioica. Nature 2004; 431(7004):67-71.
24. Duboule D, Boncinelli E, DeRobertis E et al. An update of mouse and human HOX gene nomenclature. Genomics 1990; 7(3):458-459.
25. Fritzsch G, Bohme MU, Thorndyke M et al. PCR survey of xenoturbella bocki hox genes. J Exp Zoolog B Mol Dev Evol 2008; 310(3):278-284.
26. Papillon D, Perez Y, Fasano L et al. Hox gene survey in the chaetognath spadella cephaloptera: evolutionary implications. Dev Genes Evol 2003; 213(3):142-148.
27. Aboobaker A. Hox gene evolution in nematodes: novelty conserved. Curr Opin Genet Dev 2003; 13(6):593-598.
28. Aboobaker AA, Blaxter ML. Hox gene loss during dynamic evolution of the nematode cluster. Curr Biol 2003; 13(1):37-40.
29. Wang BB, Muller-Immergluck MM, Austin J et al. A homeotic gene cluster patterns the anteroposterior body axis of C. elegans. Cell 1993; 74(1):29-42.
30. Ferrier DEK. Evolution of hox gene clusters. In: Papageorgiou S, ed. Hox Gene Expression. New York: Springer, 2007:53-67.
31. Amemiya CT, Prohaska SJ, Hill-Force A et al. The amphioxus hox cluster: characterization, comparative genomics and evolution. J Exp Zoolog B Mol Dev Evol 2008; 310(5):465-477.
32. Minguillon C, Gardenyes J, Serra E et al. No more than 14: the end of the amphioxus hox cluster. Int J Biol Sci 2005; 1(1):19-23.
33. Swalla BJ, Smith AB. Deciphering deuterostome phylogeny: molecular, morphological and palaeontological perspectives. Philos Trans R Soc Lond B Biol Sci 2008; 363(1496):1557-1568.
34. Lemons D, McGinnis W. Genomic evolution of hox gene clusters. Science 2006; 313(5795):1918-1922.
35. Amores A, Force A, Yan YL et al. Zebrafish hox clusters and vertebrate genome evolution. Science 1998; 282(5394):1711-1714.
36. Chourrout D, Delsuc F, Chourrout P et al. Minimal ProtoHox cluster inferred from bilaterian and cnidarian hox complements. Nature 2006; 442(7103):684-687.
37. Ferrier DE. Hox genes: Did the vertebrate ancestor have a Hox14? Curr Biol 2004; 14(5):R210-211.
38. Finnerty JR, Martindale MQ. The evolution of the hox cluster: insights from outgroups. Curr Opin Genet Dev 1998; 8(6):681-687.
39. Garcia-Fernandez J. Hox, ParaHox, ProtoHox: facts and guesses. Heredity 2005; 94(2):145-152.
40. Kourakis MJ, Martindale MQ. Combined-method phylogenetic analysis of hox and paraHox genes of the metazoa. J Exp Zool 2000; 288(2):175-191.
41. Ogishima S, Tanaka H. Missing link in the evolution of hox clusters. Gene 2007; 387(1-2):21-30.
42. Powers TP, Amemiya CT. Evidence for a hox14 paralog group in vertebrates. Curr Biol 2004; 14(5):R183-184.
43. Kuraku S, Takio Y, Tamura K et al. Noncanonical role of hox14 revealed by its expression patterns in lamprey and shark. Proc Natl Acad Sci USA 2008; 105(18):6679-6683.
44. Lanfear R, Bromham L. Statistical tests between competing hypotheses of hox cluster evolution. Syst Biol 2008; 57(5):1-11.
45. Schubert M, Escriva H, Xavier-Neto J et al. Amphioxus and tunicates as evolutionary model systems. Trends Ecol Evol 2006; 21(5):269-277.
46. Peterson KJ, McPeek MA, Evans DAD. Tempo and mode of early animal evolution: inferences from rocks, hox and molecular clocks. Paleobiology 2005; 31(2):36-55.
47. Pascual-Anaya J, D'Aniello S, Garcia-Fernandez J. Unexpectedly large number of conserved noncoding regions within the ancestral chordate hox cluster. Dev Genes Evol 2008; 218(11-12):591-7. Epub 2008 Sep 13.
48. Santini S, Boore JL, Meyer A. Evolutionary conservation of regulatory elements in vertebrate hox gene clusters. Genome Res 2003; 13(6A):1111-1122.

49. Prince VE, Pickett FB. Splitting pairs: the diverging fates of duplicated genes. Nat Rev Genet 2002; 3(11):827-837.
50. Mazet F, Shimeld SM. Gene duplication and divergence in the early evolution of vertebrates. Curr Opin Genet Dev 2002; 12(4):393-396.
51. Lynch M, Conery JS. The evolutionary fate and consequences of duplicate genes. Science 2000; 290(5494):1151-1155.
52. Force A, Lynch M, Pickett FB et al. Preservation of duplicate genes by complementary, degenerative mutations. Genetics 1999; 151(4):1531-1545.
53. Kondrashov FA, Rogozin IB, Wolf YI et al. Selection in the evolution of gene duplications. Genome Biol 2002; 3(2):RESEARCH0008.
54. Kondrashov FA, Kondrashov AS. Role of selection in fixation of gene duplications. J Theor Biol 2006; 239(2):141-151.
55. Van de Peer Y, Taylor JS, Braasch I et al. The ghost of selection past: rates of evolution and functional divergence of anciently duplicated genes. J Mol Evol 2001; 53(4-5):436-446.
56. Taylor JS, Raes J. Duplication and divergence: the evolution of new genes and old ideas. Annu Rev Genet 2004; 38:615-643.
57. Lynch VJ, Roth JJ, Wagner GP. Adaptive evolution of hox-gene homeodomains after cluster duplications. BMC Evol Biol 2006; 6:86.
58. McClintock JM, Carlson R, Mann DM et al. Consequences of hox gene duplication in the vertebrates: an investigation of the zebrafish hox paralogue group 1 genes. Development 2001; 128(13):2471-2484.
59. Holland PWH, GarciaFernandez J. Hox genes and chordate evolution. Dev Biol 1996; 173(2):382-395.
60. Wagner GP, Amemiya C, Ruddle F. Hox cluster duplications and the opportunity for evolutionary novelties. Proc Natl Acad Sci USA 2003; 100(25):14603-14606.
61. Budd GE. Does evolution in body patterning genes drive morphological change—or vice versa? Bioessays 1999; 21(4):326-332.
62. Hughes CL, Kaufman TC. Hox genes and the evolution of the arthropod body plan. Evol Dev 2002; 4(6):459-499.
63. Gellon G, McGinnis W. Shaping animal body plans in development and evolution by modulation of hox expression patterns. Bioessays 1998; 20(2):116-125.
64. Burke AC, Nelson CE, Morgan BA et al. Hox genes and the evolution of vertebrate axial morphology. Development 1995; 121(2):333-346.
65. Zakany J, Duboule D. The role of hox genes during vertebrate limb development. Curr Opin Genet Dev 2007; 17(4):359-366.
66. Zakany J, Kmita M, Duboule D. A dual role for hox genes in limb anterior-posterior asymmetry. Science 2004; 304(5677):1669-1672.
67. Davis AP, Witte DP, Hsieh-Li HM et al. Absence of radius and ulna in mice lacking hoxa-11 and hoxd-11. Nature 1995; 375(6534):791-795.
68. Deschamps J. Developmental biology. Hox genes in the limb: a play in two acts. Science 2004; 304(5677):1610-1611.
69. Galis F, Kundrat M, Metz JA. Hox genes, digit identities and the theropod/bird transition. J Exp Zoolog B Mol Dev Evol 2005; 304(3):198-205.
70. Gerhart J, Lowe C, Kirschner M. Hemichordates and the origin of chordates. Curr Opin Genet Dev 2005; 15(4):461-467.
71. Smith AB. Deuterostomes in a twist: the origins of a radical new body plan. Evol Dev 2008; 10(4):493-503.
72. Shen WF, Montgomery JC, Rozenfeld S et al. AbdB-like hox proteins stabilize DNA binding by the Meis1 homeodomain proteins. Mol Cell Biol 1997; 17(11):6448-6458.
73. Campos PR, de Oliveira VM, Wagner GP et al. Gene phylogenies and protein-protein interactions: possible artifacts resulting from shared protein interaction partners. J Theor Biol 2004; 231(2):197-202.
74. Welch JJ, Waxman D. Calculating independent contrasts for the comparative study of substitution rates. J Theor Biol 2008; 251(4):667-678.
75. Pagel M, Venditti C, Meade A. Large punctuational contribution of speciation to evolutionary divergence at the molecular level. Science 2006; 314(5796):119-121.
76. Xiang QY, Zhang WH, Ricklefs RE et al. Regional differences in rates of plant speciation and molecular evolution: a comparison between Eastern Asia and Eastern North America. Evolution 2004; 58(10):2175-2184.
77. Bromham L, Woolfit M, Lee MS et al. Testing the relationship between morphological and molecular rates of change along phylogenies. Evolution Int J Org Evolution 2002; 56(10):1921-1930.

78. Davies TJ, Savolainen V. Neutral theory, phylogenies and the relationship between phenotypic change and evolutionary rates. Evolution Int J Org Evolution 2006; 60(3):476-483.
79. Goh CS, Bogan AA, Joachimiak M et al. Co-evolution of proteins with their interaction partners. J Mol Biol 2000; 299(2):283-293.
80. Chiu CH, Amemiya C, Dewar K et al. Molecular evolution of the HoxA cluster in the three major gnathostome lineages. Proc Natl Acad Sci USA 2002; 99(8):5492-5497.
81. Chiu CH, Dewar K, Wagner GP et al. Bichir HoxA cluster sequence reveals surprising trends in ray-finned fish genomic evolution. Genome Res 2004; 14(1):11-17.

Section III
Biological Function

CHAPTER 9

Hox Genes and the Body Plans of Chelicerates and Pycnogonids

Wim G.M. Damen*

Abstract

Hox genes are found in all metazoan phyla and are involved in specifying identity along the anterior-posterior body axis. In arthropods, ten different classes of Hox genes can be distinguished, which are expressed in a typical staggered array along the anterior-posterior axis of the embryo in characteristically stable domains. These features have been used to align segments between different arthropod groups and in this way have contributed to solving longstanding zoological questions. In this chapter I summarize Hox gene data from chelicerates, including the enigmatic pycnogonids (sea spiders) and how these data have helped us to understand the body plans of different arthropod taxa.

Arthropods, Mandibulates vs Chelicerates

There are four major extant arthropod classes: insects, crustaceans, myriapods and chelicerates. In addition, there are the extinct trilobites.[1] Traditionally, insects and myriapods have been grouped as sister groups into the Tracheata, but molecular phylogenies suggest that the crustaceans are the sister taxon to the insects.[2] Together these three classes form the Mandibulata. The fourth arthropod class, the chelicerates are considered as a basally branching monophyletic taxon. Chelicerates have a distinct body plan that consists of two tagmata, a prosoma and an opisthosoma and have fossil representatives as early as the Cambrian. Spiders, scorpions, mites, ticks, horseshoe crabs, as well as other less familiar groups like pseudoscorpions, solifugids and most likely also the pycnogonids (sea spiders)(see also below) belong to the chelicerates.

Head segmentation in chelicerates has long been controversial.[3,4] Mandibulate arthropods exhibit a subdivision of the head region that allows easy alignment amongst the groups (Fig. 1). The antennal segment is the first appendage-bearing segment and is innervated by the ganglia of the deutocerebrum, the commissures of which are pre-oral. The next segment carries the second antenna in crustaceans, but is limbless in myriapods and insects and is called intercalary segment in these groups. The ganglia of the tritocerebrum are associated with this segment. The tritocerebral commissures are mostly post-oral. The following three segments in the mandibulates are the gnathal segments, which bear the mandibles, maxillae and second maxillae/labium, respectively (Fig. 1). Chelicerates, however, were thought to be missing the segment corresponding to the first antennal segment in the mandibulates (traditional model, Fig. 1), on the basis of analyses of the brain ganglia and the innervation of the first pair of appendages (the cheliceres). The ganglia of this cheliceral segment mostly possess post-oral commissures, similar to the tritocerebral ganglia in mandibulates. In the traditional model the cheliceral segment therefore is considered to be the

*Wim G.M. Damen—Department of Genetics, Friedrich-Schiller-University Jena, Jena, Germany. Email: wim.damen@uni-jena.de

Hox Genes: Studies from the 20th to the 21st Century, edited by Jean S. Deutsch.

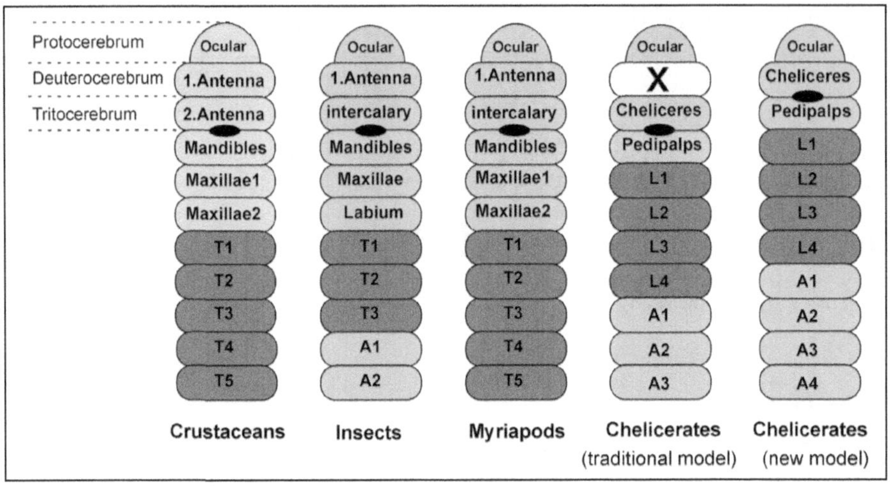

Figure 1. Head segmentation in arthropod groups. The segments are labeled according to the conventions in the respective taxa; leg-bearing segments are in dark grey. The brain parts (protocerebrum, deuterocerebrum, tritocerebrum) innervating the respective segments are indicated on the left and position of the mouth is indicated by a black oval. The alignment of the chelicerate segments is presented both as the "traditional" model and as the "new" model. In the traditional model, the presumed missing deuterocerebral segment is marked with an "X". After Damen et al[9] and Telford and Thomas.[6]

segment of the tritocerebrum and thus is aligned with the second antennal/intercalary segment of the mandibulates (Fig. 1).[1] A consequence of this model is that there would be no deuterocerebral/antennal segment in chelicerates and thus this segment has been hypothesized to have been lost in the lineage to the chelicerates (Fig. 1).

First I will give an overview of Hox genes in chelicerates. Then I will review how data on chelicerate Hox genes has helped us to understand head segmentation in arthropods. Data on Hox gene expression did not support the traditional model's alignment of segments between chelicerates and mandibulates, but suggested another model. A reinvestigation of the morphology of the nervous system has supported this new model and weakened the main arguments used in favor of the traditional model. Finally, I will discuss recent work on the pycnogonids (sea spiders) and how Hox gene analysis has provided a better understanding their enigmatic body plan.

Chelicerate Hox Genes

Hox genes have been identified in a number of different chelicerates, like the horseshoe crab *Limulus polyphemus*,[5] the mite *Archegozetes longisetosus*,[6-8] the spiders *Achaearanea tepidariorum* and *Cupiennius salei*[9-13] and the sea spiders *Endeis spinosa* and *Nymphon gracile*.[14,15] Ten different classes of Hox genes can be distinguished in chelicerates (Fig. 2).[16] Chelicerate Hox genes have been most extensively studied in the spider *Cupiennius salei*, where at least one gene for each of the ten classes of Hox genes has been identified.[13] In myriapods and crustaceans representatives of these ten classes of Hox genes have also been identified.[16-18] In the model organism *Drosophila melanogaster* (insect) only eight of the Hox genes have a homeotic function and are expressed in typical staggered domains along the anterior-posterior axis. Two Hox genes, *Hox3* and *fushi tarazu* (*ftz*), have lost their homeotic function in the lineage leading to the insects and these Hox genes acquired new functions. The insect *zerknült* (*zen*) gene is a derived *Hox3* gene that is involved in the specification of extraembryonic tissue in insects, while the *fushi tarazu* gene acts as a pair rule gene in *Drosophila*.[19]

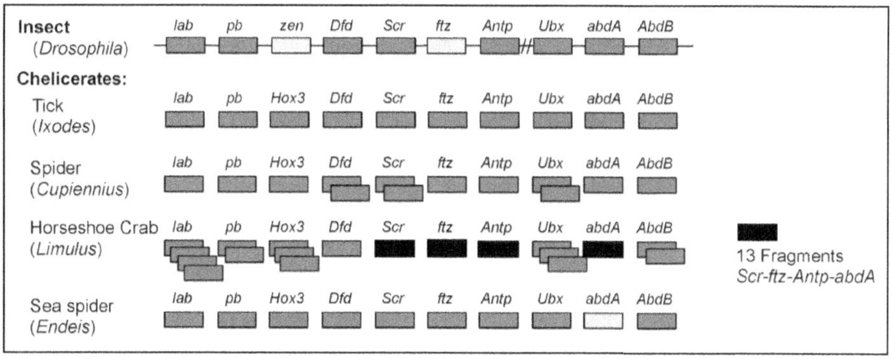

Figure 2. The ten arthropod Hox genes. Insects have a single Hox cluster; *zerknüllt* (*zen*) and *fushi tarazu* (*ftz*) no longer act as a homeotic gene in insects. In *Drosophila melanogaster* the cluster is split into the Antennapedia and the Bithorax Complexes (split marked by "//"). In chelicerates and other arthropods all ten classes of Hox genes are expressed in a Hox-like domain along the anterior-posterior body axis. The genome of the tick *Ixodes scapularis* contains a single copy of each of the ten Hox genes. PCR based screens in the spider *Cupiennius salei* recovered two copies for at least three Hox genes.[13] In the sea spider *Endeis spinosa* nine Hox genes have been recovered; no *abdA* gene has been found in *Endeis*, but a derived sequence for an *abdA* gene has been found in another sea spider, *Nymphon gracile*.[14] For the horseshoe crab *Limulus polyphemus* only sequences of short PCR fragments are available;[5] 28 different Hox-homeobox fragments have been recovered and one to four fragments can be assigned to a particular Hox cognate. Note that 13 fragments belong to the *Scr-ftz-Antp-abdA* sequences, but due to the short sequence information these fragments cannot be assigned more precisely. Data on *Cupiennius*, *Limulus* and *Endeis* are based on PCR experiments; the data represent the minimal number of Hox genes present in these species; additional Hox genes may be present.

An interesting characteristic of chelicerate Hox genes is, that at least some of them are present as duplicated copies. PCR screens in the horseshoe crab *Limulus* recovered one to four different PCR fragments of particular Hox genes, suggesting that up to four Hox clusters may be present.[5] In the spider *Achaearanea* at least one Hox gene (*Deformed*) is present as two copies and in the spider *Cupiennius* at least three Hox genes (*Deformed*, *Sex comb reduced*, *Ultrabithorax*) are present as duplicates.[9,13] However, in the sequenced genome of the tick *Ixodes scapularis* (NCBI Trace Archive v4.1) only a single copy of each of the ten Hox genes could be found (own unpublished observations). This suggests that either the duplicated genes in the horseshoe crab and the spiders are due to independent duplication events, or that there was a secondary loss of the duplicated genes in the tick. Nonetheless, the chelicerates are the only bilaterian group apart from the vertebrates for which evidence exits for duplications of multiple Hox genes or even possibly the whole cluster.[13] As PCR screens such as those done in the horseshoe crab and the spiders only allow the minimal number of Hox genes present to be determined, further research is required to establish the extent and character of these duplication events. The presence of several duplicated Hox genes or even a duplicated Hox cluster in chelicerates raises new questions on the role of Hox genes in body plan specification and even may provide insights into the evolution of duplicated Hox clusters in vertebrates.

Chelicerate Hox Genes and the Chelicerate vs Mandibulate Body Plan

Hox genes are expressed in a typical staggered array and exhibit a colinearity, which means that there is a relation between the localization of a gene in the Hox cluster and the domain of expression along the anterior-posterior axis of the embryo (e.g., Fig. 3). Genes located at the 3′ end of the cluster have more anterior expression boundaries while genes more to the 5′ end of the cluster have a more posterior expression boundary. These domains of expression along the anterior-posterior

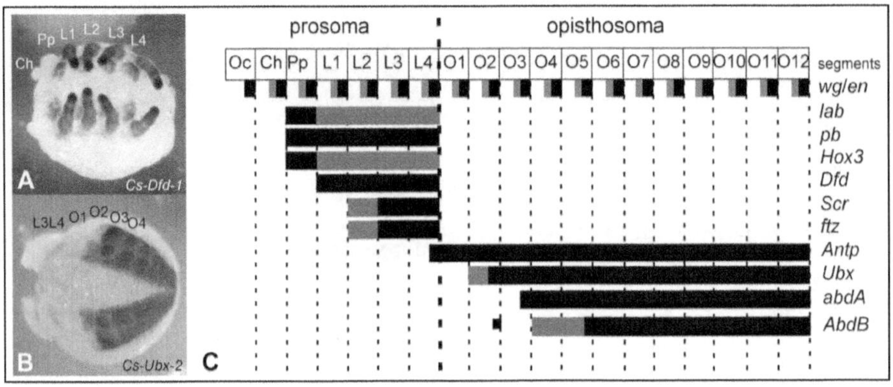

Figure 3. Expression of Hox genes in the spider *Cupiennius salei*. (A,B) Two examples of the expression of Hox genes in the spider. A) *Deformed-1* (*Cs-Dfd-1*) is expressed exclusively in the four walking leg segments. B) *Ultrabithorax-2* (*Cs-Ubx-2*) is expressed in the 2nd opisthosomal segment (O2) and the more posterior segments. C) Schematic representation of the expression domains of Hox genes in the spider *Cupiennius salei*. Bars represent the expression domains, black: strong expression, grey: weak expression in the case of *lab*, *Hox3*, *Scr* and *ftz*. The grey bar for *Ubx* indicates the more anterior expression of *Ubx-1* compared to *Ubx-2*, while the grey bar for *AbdB* represents an early expression of *AbdB* that is gone at later stages. wg/en represents the *wingless* (grey) and *engrailed* (black) expression that marks the parasegment boundary. Abbreviations: Oc: ocular or preantennal segment; Ch: cheliceral segment; Pp: Pedipalpal segment; L1-4: Walking leg segment 1-4; O1-12: opisthosomal segment 1-12. Panel (C) is slightly adapted with permission from Schwager et al[13] (©2007 Schwager et al).

axis are relatively stable. Two pioneering studies on Hox genes in a mite[6] and in a spider[9] used these features to align segments among different arthropod groups and showed the utility of Hox genes for our understanding the body plans of chelicerates versus mandibulates.

As pointed out above, establishing the homology of head segments between mandibulates and chelicerates is problematic and using the position of commissures to align segments results in a gap, which led to the traditional model in which the deuterocerebral segment is hypothesized to be missing in chelicerates (Fig. 1). Does Hox gene expression provide support for a missing deuterocerebral segment? The first indisputable appendage bearing segment in mandibulates, the antennal segment, does not express any Hox gene (summarized in ref. 16). The most anterior expressed Hox gene is *labial* in the second antenna/intercalary segment just posterior to the antennal segment. According to the traditional model, the cheliceral segment of chelicerates corresponds to this second antenna/intercalary segment in mandibulates (Fig. 1). The traditional model therefore predicts that *labial* would be expressed in this segment, but the cheliceral segment does not express any Hox gene. Instead the most anterior expression of a Hox gene is *labial* expression in the next segment, the pedipalpal segment (Fig. 3). Expression of the Hox gene *labial* thus does not support the traditional model, but suggests that the pedipalpal segment of chelicerates should be aligned with the second antenna/intercalary segment of mandibulates (Fig. 4). This inference is supported by the expression of *Deformed*. The anterior border of *Deformed* expression is one segment posterior to the *labial* expression, which is in the mandibular segment in mandibulates, while it is in the first leg segment in chelicerates (Fig. 4).[6,9] The most parsimonious explanation of the observed segmental pattern of Hox gene expression therefore is that the cheliceral segment should be aligned with the first antennal segment of the mandibulates and thus is the deuterocerebral segment. Consequentially there is no need to propose a missing deuterocerebral segment for chelicerates (Fig. 1). A wealth of Hox gene expression data that appeared after these two publications supports this view.[16-18]

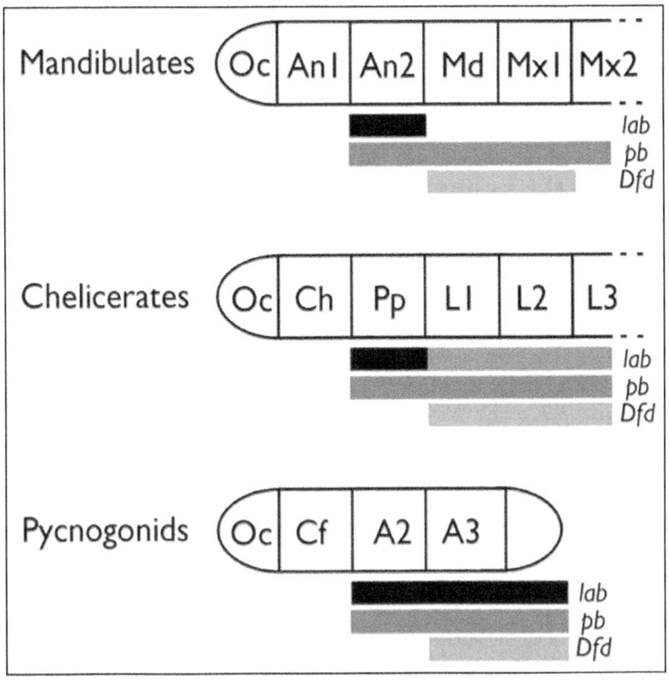

Figure 4. Expression of Hox genes in the anterior segments. Shown is a schematic representation of the expression of *labial* (*lab*), *proboscipedia* (*pb*) and *Deformed* (*Dfd*) in Mandibulates, Chelicerates and Pycnogonids. After Jager et al,[15] Damen et al[9] and Telford and Thomas.[6] Abbreviations: Oc: ocular or preantennal segment; An1: first antennal segment; An2: second antennal segment (or intercalary in insects and myriapods); Md: mandibular segment; Mx: maxillary segment; Ch: cheliceral segment; Pp: Pedipalpal segment; L1-4: Walking leg segment 1-4; Cf: Chelifore segment; A2-3: 2nd and 3rd larval appendage bearing segment.

It should be noted that the protocerebrum, the anterior part of the arthropod brain, has never been an issue in this discussion, since it does not innervate any appendage in extant arthropods. The protocerebrum is considered either to be a nonsegmented anterior structure (the acron), or it may represent the neuromere of an anterior segment that might have borne an appendage in an early ancestor (see also next section on sea spiders). The brain of all extant arthropods thus consists of a protocerebrum, deuterocerebrum and a tritocerebrum.

The new model is, however, in conflict with the assumed position of the ganglia and the commissures, as it implies that the deuterocerebral commissures are post-oral in chelicerates and not pre-oral like in mandibulates. As pointed out above, the post-oral position of the commissures of the ganglia innervating the cheliceres provides the strongest argument in favor of the traditional model, as their organization is similar to the tritocerebral ganglia of mandibulates. This issue, however, has long been controversial as there are several chelicerates that have cheliceral commissures in a pre-oral position and this feature is not exclusively post-oral. Recent detailed studies of the brain in the horseshoe crab show that the cheliceral commissure in the horseshoe crab resides principally in a pre-oral position. This would be unique for a tritocerebral commissure.[20] Neurotransmitter expression data are in accordance with this view.[21] Furthermore, neither in the embryo, nor in the larvae or adult could remnants of a missing segment be found. The position of the commissures alone is not a good character for concluding segmental homologies of the arthropod head segments.

The work on Hox genes in the spider[9] and the mite[6] thus helped to invigorate the long-standing discussion on the controversial issue of the head segments in chelicerates. But it must be noted that the "new model" as supported by the Hox gene expression data in fact is not that new. A model claiming this homology was presented long ago,[22,23] but then was overshadowed by what became the traditional model.

Hox Genes and the Enigmatic Sea Spider Body Plan

Sea spiders (pycnogonids) are an enigmatic arthropod group. Even the position of pycnogonids within the arthropods is debated; molecular and total-evidence data are not consistent on their phylogenetic position.[24,25] They are commonly considered either as a sister group to the chelicerates, or as a sister group to all other arthropods.[24,25] Morphologically, pycnogonids have few characters that can be used to place them. The most convincing (or maybe the only convincing)[24] synapomorphy is the first pair of appendages, the chelifores, which are pincer-bearing appendages that are assumed to be related to the cheliceres, the pincered fangs of chelicerates. This raises the question of whether the chelifores of pycnogonids and the cheliceres of chelicerates are indeed the appendages of a homologous segment and thus homologous appendages.

A recent study on the neuroanatomy of a sea spider challenged the homology of pycnogonid chelifores and chelicerate cheliceres.[26] This study suggested that the chelifores of pycnogonids are innervated by the protocerebrum and not by the deuterocerebrum like the cheliceres of chelicerates. A direct consequence of this association of chelifores and cheliceres with different parts of the brain would be that these appendages are not homologous, but are in fact derived from different segments. The protocerebrum, the anterior-most part of the arthropod brain does not innervate an appendage in any other extant arthropod. However, there is fossil evidence that early arthropods had a pair of large appendages at the anterior of the head, the so-called great appendage and it even has been suggested that these great appendages were innervated by the protocerebrum.[27,28] An exciting implication of this would be that pycnogonids have retained an organization of the head with appendages on a protocerebral segment, which has been lost in all other living arthropods.

However, Hox gene expression data do not support this idea. As discussed in the previous section, Hox genes have relatively stable expression domains along the anterior-posterior axis of arthropods and this has been used by Jager et al[15] to test the proposed association of pycnogonid chelifores with the protocerebrum. In the larvae of the sea spider *Endeis spinosa* the anterior-most expression of *labial* and *proboscipedia* is in the second larval appendage, while the anterior-most expression of *Deformed* is in the third larval appendage (Fig. 4).[15] As noted previously the most anterior segment that expresses Hox genes in all other arthropods is the tritocerebral segment that expresses *labial* and *proboscipedia*, while the anterior border of *Deformed* is in the next segment (Fig. 1 and Fig. 4).[16] The expression domains of these Hox genes thus suggest that the segment bearing the second larval appendage in the sea spider is likely to be the tritocerebral segment. The segment with the chelifores is just anterior to this segment and just like the cheliceral segment does not express any Hox gene, which is consistent with a deuterocerebral identity of this segment (Fig. 4). The Hox expression data thus strongly suggests that the chelifore segment is the deuterocerebral segment of pycnogonids and builds a strong argument for the homology of the chelifores with the chelicerate cheliceres and the mandibulate first antennae. A further consequence of this segment alignment, however, is that the chelifores are not actually associated with the protocerebrum and thus there is no support for an extant arthropod with protocerebral great appendages.

Since the Hox expression data[15] are not consistent with the conclusion of the neuroanatomical analyses[26] one is presented with a clear conflict. A reinterpretation of the neuroanatomy data however could solve the problem.[3,15] One source of the conflict could be caused by the possibility that the chelifore ganglia initially form in a posterior position and only later migrate and come to lie in a position just anterior and in close proximity, to the protocerebrum. A similar forward movement is seen for the cheliceral ganglia in arachnids.[29] In both pycnogonids and chelicerates there is thus a tendency towards an anterior migration of ganglia, but this pattern of neural morphogenesis may be variable within various clades. As mentioned above, there are several chelicerates that have cheliceral commissures in a pre-oral position and not exclusively in a post-oral position.[9] Similar variance has been described for sea spider species; where in some species the ganglia of the second larval appendage pair have a post-oral commissure, whereas in other species they have a pre-oral commissure.[15] Due to the relative migration of ganglia during embryogenesis there are variations in the position of the commissures and it might be hard to recognize the ganglia exactly based on these characters.

Conclusion

The two examples presented here show the power of Hox gene expression data in discussions on arthropod body plans, particularly in the controversial discussion of the head segments of chelicerates and pycnogonids. Indeed the anterior expression border of the anterior Hox genes like *labial* and *Deformed* as used in these studies is stable and appear to be especially useful as markers. However, it should also be noted that there are examples of posterior Hox genes where the anterior expression margin varies within a group, as is the case in crustaceans. Changes in the expression pattern of *Ultrabithorax* in different crustaceans correlate well with the modification of their anterior thoracic limbs into maxillipeds (feeding appendages) and a morphological change is associated with changes in Hox gene expression.[30] Such variation, however, has never been observed for *labial* and *Deformed*, the anterior Hox genes used in the studies described above.

Finally, as molecules and morphology both describe the same history of an animal or animal group, studies of each should in this give the same outcome. Any discrepancy between conclusions therefore should reinforce the discussion of the arguments and stimulate if necessary new experiments. In retrorespect, the Hox gene studies in the spider[9] and the mite[6] did this and caused a revitalization of the debate on the homology of anterior segments in arthropods.[20] Hopefully, the new studies in sea spiders[15,26] will cause a similar discussion on the head segments of the pycnogonids and the origin of the arthropod head.[28]

Acknowledgements

I thank the members of my lab for discussions and Alistair McGregor for critical reading of the manuscript. The work in my lab is supported in part by the DFG via SFB 572 of the University of Cologne and by the European Union via the Marie Curie Research and Training Network ZOONET (MRTN-CT-2004-005624).

References

1. Westheide W, Rieger R. Spezielle Zoologie Teil 1: Einzeller und Wirbellose Tiere, 1st ed. Stuttgart: Gustav Fischer Verlag, 1996.
2. Friedrich M, Tautz D. Ribosomal DNA phylogeny of the major extant arthropod classes and the evolution of myriapods. Nature 1995; 376:165-167.
3. Bitsch J, Bitsch C. The segmental organization of the head region in Chelicerata: a critical review of recent studies and hypotheses. Acta Zoologica (Stockholm) 2007; 88:317-335.
4. Scholtz G, Edgecombe GD. The evolution of arthropod heads: reconciling morphological, developmental and palaeontological evidence. Dev Genes Evol 2006; 216:395-415.
5. Cartwright P, Dick M, Buss LW. HOM/Hox type homeoboxes in the chelicerate Limulus polyphemus. Mol Phyl Evol 1993; 2:185-192.
6. Telford MJ, Thomas RH. Expression of homeobox genes shows chelicerate arthropods retain their deuterocerebral segment. Proc Natl Acad Sci USA 1998; 95:10671-10675.
7. Telford MJ, Thomas RH. Of mites and zen: expression studies in a chelicerate arthropod confirm zen is a divergent Hox gene. Dev Genes Evol 1998; 208:591-594.
8. Telford MJ. Evidence for the derivation of the Drosophila fushi tarazu gene from a Hox gene orthologous to lophotrochozoan Lox5. Curr Biol 2000; 10:349-253.
9. Damen WGM, Hausdorf M, Seyfarth E-A et al. A conserved mode of head segmentation in arthropods revealed by the expression pattern of Hox genes in a spider. Proc Natl Acad Sci USA 1998; 95:10665-10670.
10. Abzhanov A, Popadic A, Kaufman TC. Chelicerate Hox genes and the homology of arthropod segments. Evol Dev 1999; 1:77-89.
11. Damen WGM, Tautz D. A HOX class 3 orthologue from the spider Cupiennius salei is expressed in a Hox-gene-like fashion. Dev Genes Evol 1998; 208:586-590.

12. Damen WGM, Janssen R, Prpic N-M. Pair rule gene orthologs in spider segmentation. Evol Dev 2005; 7:618-628.
13. Schwager EE, Schoppmeier M, Pechmann M et al. Duplicated Hox genes in the spider Cupiennius salei. Front Zool 2007; 4:10.
14. Manuel M, Jager M, Murienne J et al. Hox genes in sea spiders (Pycnogonida) and the homology of arthropod head segments. Dev Genes Evol 2006; 216:481-491.
15. Jager M, Murienne J, Clabaut C et al. Homology of arthropod anterior appendages revealed by Hox gene expression in a sea spider. Nature 2006; 441:506-508.
16. Hughes CL, Kaufman TC. Hox genes and the evolution of the arthropod body plan. Evol Dev 2002; 4:459-499.
17. Hughes CL, Kaufman TC. Exploring the myriapod body plan: expression patterns of the ten Hox genes in a centipede. Development 2002; 19:1225-1238.
18. Janssen R, Damen WGM. The ten Hox genes of the millipede Glomeris marginata. Dev Genes Evol 2006; 216:451-465.
19. Damen WGM. fushi tarazu: A Hox gene changes its role. BioEssays 2002; 24:992-995.
20. Mittmann B, Scholtz G. Development of the nervous system in the "head" of Limulus polyphemus (Chelicerata: Xiphosura): morphological evidence for a correspondence between the segments of the chelicerae and of the (first) antennae of Mandibulata. Dev Genes Evol 2003; 213:9-17.
21. Harzsch S, Wildt M, Battelle B et al. Immunohistochemical localization of neurotransmitters in the nervous system of larval Limulus polyphemus (Chelicerata, Xiphosura): evidence for a conserved protocerebral architecture in Euarthropoda. Arthropod Struct Dev 2005; 34:327-342.
22. Viallanes MH. Etudes histologiques et organologiques sur les centres nerveux et les organs de sens des animaux articules. Ann Sci Nat Serv VII Zool 1893; 14:405-456.
23. Goodrich ES. On the relation of the arthropod head to the annelid prostomium. Q J Microsc Sci 1897; 40:147-268.
24. Dunlop JA, Arango CP. Pycnogonid affinities: a review. Journal of Zoology, Systematics and Evolutionary Research 2005; 43:8-21.
25. Mallatt J, Giribet G. Further use of nearly complete 28S and 18S rRNA genes to classify Ecdysozoa: 37 more arthropods and a kinorhynch. Mol Phylogenet Evol 2006; 40:772-794.
26. Maxmen A, Browne WE, Martindale MQ et al. Neuroanatomy of sea spiders implies an appendicular origin of the protocerebral segment. Nature 2005; 437:1144-1148.
27. Budd GE. A palaeontological solution to the arthropod head problem. Nature 2002; 417:271-275.
28. Budd GE, Telford MJ. Along came a sea spider. Nature 2005; 437:1099-1102.
29. Weygoldt P. Ontogeny of the arachnid central nervous system. In: Barth FG, ed. Neurobiology of Arachnids. Heidelberg Springer Verlag 1985:20-37.
30. Averof M, Patel NH. Crustacean appendage evolution associated with changes in Hox gene expression. Nature 1997; 388:682-686.

Chapter 10

Hox3/zen and the Evolution of Extraembryonic Epithelia in Insects

Urs Schmidt-Ott,* Ab. Matteen Rafiqi and Steffen Lemke

Abstract

Insects have undergone dramatic evolutionary changes in extraembryonic development, which correlate with changes in the expression of the class-3 Hox gene *zen*. Here, we review the evolution of this gene in insects and point out how changes in *zen* expression may have affected extraembryonic development at the morphological and the genetic level.

Introduction

During the early radiation of insects, *Hox3* abandoned its ancestral role in specifying segmental identity along the anteroposterior axis of the embryo and acquired a new role in extraembryonic tissue.[1] This evolutionary transition may have occurred in the stem lineage of modern Pterygota (winged insects), as indicated by expression data and protein comparisons. Canonical Hox proteins share a hexapeptide or YPWM motif,[1] which enables them to interact with the Hox-cofactor Exd/Pbx.[2,3] In the context of overlapping Hox gene expression patterns, e.g., during axis-specification, this interaction is important because it contributes indirectly to the DNA-binding specificity of individual Hox proteins.[4] The *Hox3* gene of the apterygotan firebrat *Thermobia* is expressed in a nested arrangement with other Hox genes in the prospective gnathocephalon and in the growth zone and encodes a protein with the YPWM motif.[5,6] In contrast, *Hox3* genes of Pterygota (named *zen* after their *Drosophila* prototype)[7] are expressed in extraembryonic tissue, lack expression in the germband and encode proteins without YPWM motif.[6] The present chapter is devoted to the evolutionary history of extraembryonic *zen* expression in insects. We cover data from five orders including the silverfish *Thermobia* (Thysanura), the grasshopper *Schistocerca* (Orthoptera), the bug *Oncopeltus* (Hemiptera), the beetle *Tribolium* (Coleoptera) and various fly species (Diptera). Yet, the focus of this review is on Diptera as this is the only insect order for which data on the expression and function of *zen* are available from multiple distantly related species. Even in this order, some relevant traits have been poorly sampled. Despite the 'patchiness' of the data matrix, we hope that as current working hypotheses our phylogentic inferences will help to advance the field. To set the stage, we briefly review morphological aspects of extraembryonic development. In the following sections, we focus on variants of *zen* expression and their evolutionary significance. We propose that changes in *zen* expression underlie major reorganizations in extraembryonic development of pterygote insects.

*Corresponding Author: Urs Schmidt-Ott—University of Chicago, Department of Organismal Biology and Anatomy, 920 E. 58th Street, Chicago, IL 60637. Email: uschmid@uchicago.edu

Hox Genes: Studies from the 20th to the 21st Century, edited by Jean S. Deutsch.

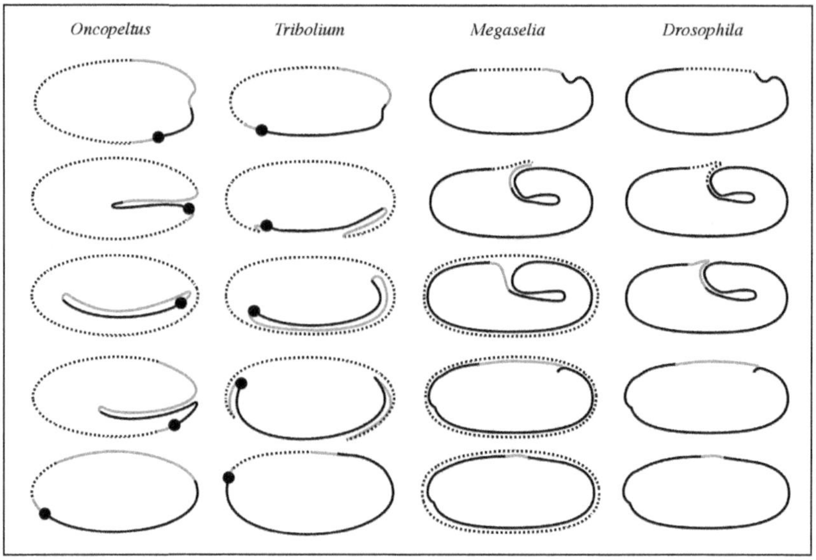

Figure 1. Schematic illustrations of extraembryonic developmental trajectories. Embryonic (black line), amniotic (grey) and serosal tissues (broken line) are indicated at consecutive developmental stages. Sketches are based on *Oncopeltus*,[9,10] *Tribolium*,[23] *Megaselia*[18] and *Drosophila*.[19] A filled circle indicates the position of the head in *Oncopeltus* and *Tribolium* embryos.

Setting the Stage: Morphological Evolution of Extraembryonic Development

Extraembryonic development of insects begins at the blastoderm stage, when a single cell layer forms around the yolk. At this stage, a portion of anterior or dorsal blastoderm—depending on the species—is specified to become the serosa, an epithelium underneath the eggshell, which secretes a cuticle.[8] Typically, serosa formation occurs by invagination of the posterior blastoderm (e.g., *Oncopeltus*, Fig. 1), or by closure over the ventral blastoderm (e.g., *Tribolium*, Fig. 1). In both cases, the internalized blastoderm pinches off from the serosa as a flattened hollow body, which is composed of a thin cell layer (amnion) and the gastrulating embryo. The fold through which part of the blastoderm is internalized is called 'amnioserosal fold' and the yolk-free space between the amnion and the embryo is called 'amniotic cavity'. Later in development, while the flanks of the embryo close along the dorsal midline (dorsal closure), the process of serosa and amnion formation is reversed. The serosa and the amnion generate a continuous serosa-amnion epithelium ('serosa-amnion fusion'), which retracts towards the anterior or dorsal side of the egg where both epithelia are resorbed by the yolk. This process ruptures the amniotic cavity and regenerates an opening in the extraembryonic epithelia ('serosal window'). In hemimetabolous insects (lower Pterygota[a]), serosa-amnion fusion seems to be widely conserved, as it is required for realigning the inverted anteroposterior (A-P) axis of the early embryo with the A-P axis of the egg (katatrepsis).[9,10] In holometabolous insects (higher Pterygota), serosa-amnion fusion has been observed as well, but major variants of extraembryonic development have been found in all the large orders, including Hymenoptera, Coleoptera, Lepidoptera and Diptera.[8,11-13] Here we limit the discussion to variants within the dipteran order. In lower Diptera (noncyclorrhaphan flies), serosa and amnion

[a] Throughout this review, we use the term 'lower' in conjunction with a taxon name to designate basal branches of monophyletic taxa. Conversely, the term 'higher' is used when the intention is to exclude species on basal branches.

are generated essentially in the same way as in the beetle *Tribolium* (Fig. 1).[14] However, it seems that in many lower dipterans the two mature epithelia fail to fuse again and persist or degenerate independently, while dorsal closure proceeds.[15-17] More radical departures from the ancestral trajectory have been reported for higher dipterans (Cyclorrhapha). Lower cyclorrhaphan flies such as the phorid fly *Megaselia abdita* or the syrphid fly *Episyrphus balteatus* initiate the formation of an amnioserosal fold, but while their serosa expands ventrally, their amnion disjoins from the leading edge of the serosa and grows over the dorsal side of the yolk sac (e.g., *Megaselia*, Fig. 1).[18] In these species, serosa-amnion fusion does not occur and only the amnion is resorbed by the yolk. Finally, in higher cyclorrhaphan flies, while a rudimentary amnioserosal fold is transiently visible at the morphological level, the extraembryonic tissue does not disjoin at the edge of this fold and does not expand over the germband (e.g., *Drosophila*, Fig. 1). Instead, the extraembryonic anlage gives rise to a single epithelium, called amnioserosa, which is later resorbed by the yolk,[19,20] like the dorsal amnion in lower Cyclorrhapha. Taken together, the data suggest that extraembryonic development in the dipteran lineage evolved in three distinct steps. First, mature serosal and amniotic epithelia failed to fuse and retract as a continuous serosa-amnion epithelium during the process of dorsal closure. Second, ventral closure of the amnion was suppressed and substituted by closure of the amnion over the dorsal yolk sac. Third, serosa and dorsal amnion were transformed into an amnioserosa. All these evolutionary transitions seem to correlate with changes in the activity pattern of *zen*, which will be reviewed in the next section.

Variants of *zen* Expression and Function in Insects and Possible Morphological Correlates

Variants of *zen* expression and function in insects (Fig. 2) suggest that this gene played an important role in the evolution of extraembryonic tissue. The apterygotan insect *Thermobia* expresses *Hox3* not only in the embryo (see Introduction) but also in the mature (completed) amnion.[5,b] Hemimetabolous Pterygota (e.g., *Schistocerca*, *Oncopeltus*) exhibit *zen* expression transiently during blastoderm formation and subsequently in the serosa and parts of the amnion.[9,21,22] In the grasshopper *Schistocerca gregaria*, extraembryonic *zen* expression begins in the leading edge of the developing serosa ('necklace cells') but in other species comparable *zen* expression has not been observed. Hence, *zen* expression in necklace cells might have evolved in the *Schistocerca* lineage. During later stages, *Schistocerca* expresses *zen* throughout the completed serosa and in adjacent parts of the completed amnion. In the bug *Oncopeltus fasciatus*, *zen* expression has been observed throughout the completed serosa and in a rim of amniotic cells, which connect with the contracting serosa. RNA interference (RNAi) against *Oncopeltus zen* (*Of-zen*) prevents rupture of the amniotic cavity and formation of a continuous serosa-amnion epithelium, as well as katatrepsis and dorsal closure.[9,10,c] To explain this phenotype, *zen* activity in the serosa has been invoked as a regulator of global contractions,[10] but an essential complementary role of amniotic *zen* expression in this process has not been ruled out.

In holometabolous insects, serosal tissue expresses *zen* at all stages.[18,21,23-25] Amniotic *zen* expression has been reported for a beetle (*Tribolium*) but only at the site where the serosal window opens up prior to dorsal closure. In species that lack this process (e.g., the lower cyclorrhaphan flies *Megaselia* and *Episyrphus*), amniotic *zen* expression has not been observed. In *Tribolium castaneum*, *zen* is critical for both serosa specification and serosa-amnion fusion. The distinction of these functions is possible because of a recent duplication of the *zen* locus in the *Tribolium* lineage[26] and subfunctionalization of the resulting *Tribolium* paralogs, *Tc-zen1* and *Tc-zen2*.[23] Both genes are expressed throughout the serosal tissue, while only *Tc-zen2* is expressed in the

[b] Amniotic expression was observed at the torpedo stage, when extraembryonic epithelia have already formed.

[c] In *Of-zen* RNAi embryos, the serosa and the amnion form a 'serosal window pane' that fails to rupture, keeping the amniotic cavity closed and detracting the flanks of the embryo, which under these conditions close ventrally.

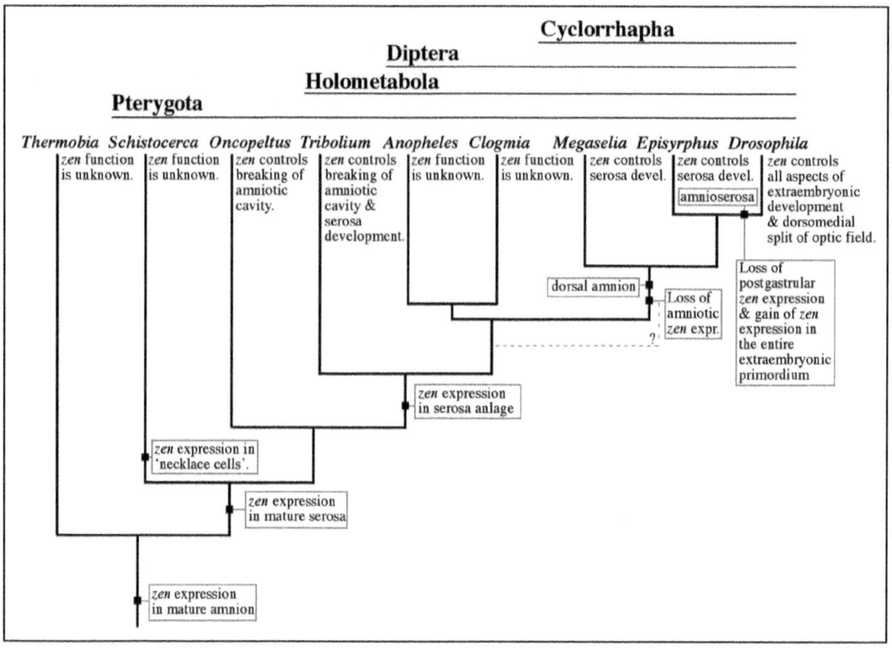

Figure 2. Evolution of extraembryonic *zen* expression and function. Phylogenetic relationships are shown according to Grimaldi and Engel.[51] For details see text. For the eggs of *Schistocerca* and diverse lower dipterans (including *Clogmia*), the presence of maternal *zen* transcripts has been documented.[21,22,25] *Anopheles, Megaselia, Episyrphus* and *Drosophila* lack maternal *zen* expression.[7,18,24,25]

amnion. RNAi against *Tc-zen1,* which is epistatic to serosal expression of *Tc-zen2,* suppresses serosa development and causes the formation of a single extraembryonic epithelium in which all cells are amnion-like. *Tc-zen2* RNAi suppresses serosa-amnion fusion and timely rupture of the amniotic cavity, indicating a possible function of amniotic *zen* expression in this process. The function of *Tc-zen1* is probably conserved across a wide range of holometabolous insects. For example, in *Megaselia abdita,* RNAi against *zen* suppresses serosa development and results in the formation of a single, amnion-like extraembryonic epithelium, which is sufficient to support germband retraction and dorsal closure,[18] just like in *Tribolium.* In summary, *zen* expression in the mature serosa has been reported for a wide range of (pterygote) insects, while *zen* expression in the serosa anlage and *zen*-dependent serosa specification might be characteristic of Holometabola. *zen* expression in the late amnion may have evolved before the radiation of Pterygota but seems to be restricted to species that undergo serosa-amnion fusion prior to katatrepsis (Hemimetabola only) and dorsal closure and we suspect that the amniotic domain is required for the formation of a continuous serosa-amnion epithelium.

Among holometabolous insects the role of *zen* in *Drosophila* stands out. Unlike the beetle *Tribolium* and the lower cyclorrhaphan flies *Megaselia* and *Episyrphus,* the higher cylorrhaphan (schizophoran) fly *Drosophila* is unable to specify any extraembryonic tissue in the absence of *zen* activity.[d] Also the expression pattern of *Drosophila zen* appears diverged. Its novel features

[d] *Drosophila melanogaster* contains two copies of the *zen* locus (*zen, zen2*), which are expressed similarly. We do not consider *zen2* separately because available mutants of *zen* also remove the activity of *zen2* and because *zen2* was shown to be dispensable for normal development.

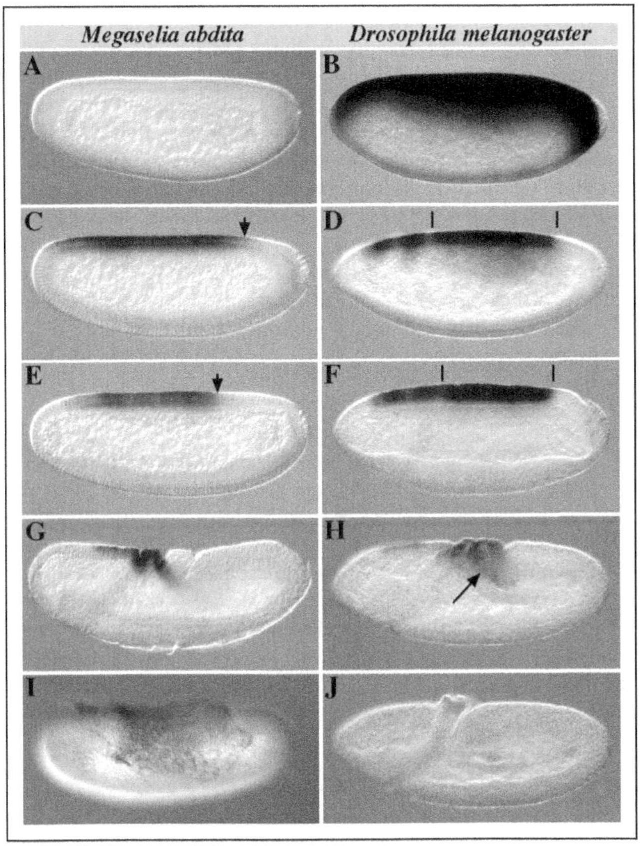

Figure 3. Expression of *Mab-zen/zen* transcripts in *Megaselia abdita* and *Drosophila melanogaster.* Matching consecutive stages of *Megaselia* (left) and *Drosophila* (right) are shown before cellularization (A,B), during cellularization (C,D), at the beginning of gastrulation (E,F), during gastrulation (G,H) and after gastrulation (I,J). Horizontal bars indicate the position of the amnioserosa anlage. Posterior shortening of the *Mab-zen* domain is indicated by arrowheads and the *zen*-positive invaginated portion of the amnioserosa is indicated by an arrow. Panels C, I and J from reference 18. Anterior is left and dorsal up.

include a broad dorsal expression domain at early blastoderm stages, a posteriorly expanded and laterally compressed expression domain of *zen* at late blastoderm stages and absence of *zen* expression after gastrulation (Fig. 3). Below, we discuss the evolutionary significance of these *Drosophila*-specific features.

Suppression of Postgastrular zen *Expression May Have Triggered the Origin of the Amnioserosa*

In holometabolous insects, serosal *zen* expression is maintained from the time of serosa specification until after the completion of the serosa epithelium.[18,23] However, in *Drosophila, zen* expression (transcript and protein) fades during stage 8, shortly after a pseudo-amnioserosal fold has formed. We found in *Megaselia* that during early gastrulation *zen* expression is still critical for the formation of distinct serosal and amniotic epithelia.[28] Suppression of *zen* expression at this stage alters the developmental trajectory of the extraembryonic epithelia: the serosa fails to expand over the germband and does not disjoin from the edge of the amnion but rather becomes part of it or disintegrates. Hence, the amnioserosa could have originated by the loss of postgastrular *zen* expression, which suppressed further serosa development while allowing the formation of a *zen*-negative dorsal amnion. This hypothesis implies that late developmental stages of the ancestral amnioserosa (after stage 8) were equivalent to a dorsal amnion, whereas early developmental stages of the ancestral amnioserosa (until stage 8) were composed of *zen*-expressing serosal and *zen*-negative amniotic tissue. However, in *Drosophila*, all cells of the early amnioserosa express *zen* (including cells that invaginate with the proctodeum and which formerly gave rise to the amnion) and hence, might best be compared with early serosa cells of lower cyclorrhaphan flies. This novel feature of extraembryonic development may have evolved after the origin of the amnioserosa because at this point in time the distinction of serosal and amniotic primordia had become obsolete. The genetic mechanisms by which *zen* gained control over all amnion-competent cells of the *Drosophila* blastoderm were probably different along the anteroposterior (A-P) and the dorsoventral (D-V) axis and will be discussed separately.

Reduction of the Amniotic Anlage along the A-P Axis

In *Megaselia*, the expression domain of *zen* in the cellular blastoderm is shortened at the posterior end. The cells, in which *zen* has been actively repressed, invaginate with the proctodeum and become part of the amnion (cf. Figs. 1 and 3). The repressor has not been identified but could be dependent on *Megaselia caudal* (*Mab-cad*), which is specifically expressed in amniotic blastoderm.[29] *Drosophila* lacks expression of *caudal* in this domain of the blastoderm and does not down-regulate *zen* there. As a result, *zen* is also expressed in cells that eventually invaginate with the proctodeum (cf. Figs. 1 and 3). These cells become part of the amnioserosa. Thus, the posterior expansion of *zen* expression in the cellular blastoderm of *Drosophila* may have caused a reduction of the amnion anlage along the A-P axis. Consistent with this hypothesis, we recently found that the injection of capped *Mab-zen* mRNA into syncytial *Megaselia* embryos represses genes with amniotic expression and causes defects in germband retraction and dorsal closure, but does not interfere with the formation of the serosa (unpublished data).

Reduction of the Amniotic Anlage along the D-V Axis

While derepression of posterior *zen* activity in the *Drosophila* lineage may account for the reduction of the amniotic anlage along the A-P axis, a different mechanism must account for the reduction of the amniotic anlage along the D-V axis, because in the cellular blastoderm of *Drosophila*, the expression domain of *zen* is narrower than in other dipterans.[24] How was this transition achieved? Below, we argue that the gain of early broad *zen* expression (spanning about 40% of the D-V perimeter) in the *Drosophila* lineage was critical for the evolution of an all-*zen*-dependent extraembryonic anlage. To make the argument, it is necessary to introduce another *Drosophila* gene, *decapentaplegic* (*dpp*).[30] This gene encodes a ligand of the bone morphogenetic protein (BMP) family. It is secreted into the space between the eggshell and the embryo and promotes pattern formation in the dorsal ectoderm in a concentration-dependent manner (reviewed in ref. 31). The dorsal blastoderm of *dpp*-deficient *Drosophila* embryos acquires a more ventral neurogenetic fate; moderate levels of *dpp* activity are sufficient to specify dorsal embryonic ectoderm (but insufficient for specifying the amnioserosa); and high-level overexpression of *dpp* converts all dorsal cells to an amnioserosa fate.[32]

The activity of *dpp* can be visualized with an antibody against the phosphorylated form of the intracellular protein Mad, which transmits the Dpp-dependent signal from the cytoplasm to the regulatory DNA sequences of target genes.[33-35] As secreted Dpp undergoes extracellular transport towards the dorsal midline, its activity range in the dorsal ectoderm is confined to a narrower portion than the distribution of *dpp* transcript (ca. 40% of the D-V perimeter) would suggest (reviewed in ref. 31). Long-range, extracellular transport of Dpp accounts for the shallow pMad gradient that is observed in the early *Drosophila* blastoderm and which spans about 20% of the D-V perimeter with peak levels at the dorsal midline. In older embryos (undergoing blastoderm cellularization), pMAD levels experience an additional and much more dramatic lateral contraction (accompanied by dorsal increase) in response to a positive feedback loop that is at least in part

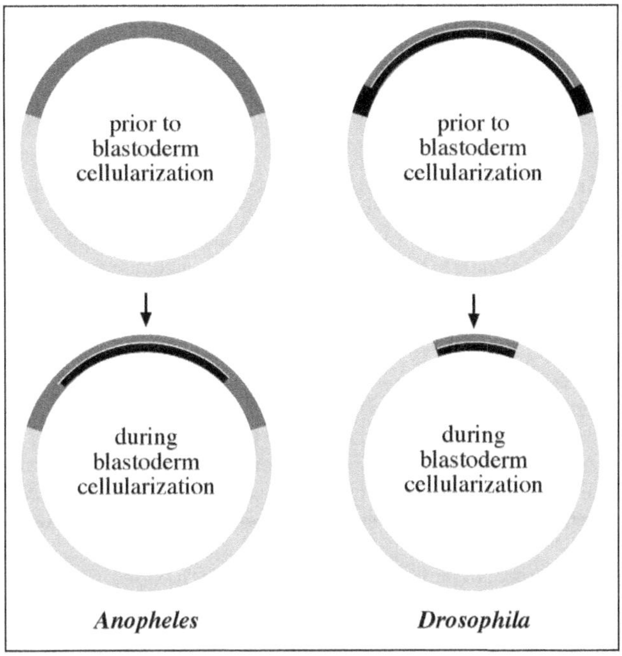

Figure 4. Evolution of the relationship of *zen* and *dpp* activities in the blastoderm. Blastoderm embryos of *Anopheles* (left) and *Drosophila* (right) are depicted as transverse sections at two consecutive stages with *dpp* activity in dark grey and *zen* activity in black. Note that in *Anopheles*, *dpp* activity extends beyond the expression domain of *zen*, while in *Drosophila*, *dpp* activity does not extend beyond the activity range of *zen*. For details and references see text.

dependent on *zen*.[36] The result of the refinement is a high-level pMad domain that now spans only about 10% of the D-V perimeter, precisely the width of the amnioserosa anlage.[e] The important point is that throughout blastoderm development, *dpp* activity of the *Drosophila* embryo does not extend beyond the expression domain of *zen*. This is different in lower dipterans.

Compared to *Drosophila*, the spatial relationship of Dpp activity and *zen* expression in the mosquito *Anopheles* appears to be inversed (Fig. 4). In *Anopheles*, the pMad domain is much broader than in *Drosophila* and extends at all blastoderm stages beyond the boundaries of the *zen* domain.[24] Provided that *Anopheles dpp* is essential for the specification of amnion-competent blastoderm, as it probably is[f], and considering that *zen* represses amnion development in amnion-competent blastoderm,[18,23] the gain of early broad *zen* expression in the *Drosophila* lineage after the origin of the amnioserosa might well have delayed amnion specification until the end of gastrulation by promoting serosa development at early stages.

[e] The evolutionary trend towards bistable Dpp activity in the late blastoderm could have gradually increased the ratio of serosal to amnion-competent blastoderm and might have prevented ventral amnion closure in the stem lineage of schizophoran flies.

[f] In *Tribolium*, the effect of *dpp* RNAi on extraembryonic development has been assessed using early markers for the amnion (*Tc-pnr*) and a dorsal sector of the serosa anlage (*Tc-doc*). *Tc-dpp* RNAi abolishes *Tc-pnr* as well as *Tc-doc* expression. However, only amnion development is suppressed, while the serosa primordium is merely reduced. In *Tribolium*, the specification of the serosa strongly depends on anterior signaling-input from a receptor tyrosine kinase ('terminal system'). In *Oncopeltus*, *dpp* RNAi suppresses the invagination of the germ rudiment, which might depend on the specification of amniotic tissue, but does not interfere with serosa development.

Expression of zen *in the Optic Field*

Drosophila zen has also an embryonic function. In wildtype embryos of *Drosophila*, regulatory genes that establish the optic field, such as *sine oculis* (*so*) and *eyes absent* (*eya*), are repressed at the dorsal midline in response to *zen*, which prevents a cyclopic phenotype.[40] However, in *Megaselia* and *Episyrphus*, *zen* RNAi does not affect the visual system. In these species, the specification of amnion-competent blastoderm under the control of *dpp* might be sufficient for splitting the optic field into left and right parts. In *Drosophila*, genes of the optic field may have acquired Zen binding sites in their *cis*-regulatory DNA. This model implies a repressor function of Zen. Alternatively, *Drosophila* genes of the optic field may have retained the ancestral, Dpp-dependent regulatory mechanism. In this case, it should be possible to rescue the phenotype of *zen*-mutant embryos in the optic field by increasing Dpp activity along the dorsal midline.

The Amnioserosa Gene-Network in Evolutionary Perspective

Above, we argued that in the *Drosophila* lineage, the loss of postgastrular *zen* expression abrogated serosa development and resulted in the formation of a single extraembryonic epithelium, while boundary conflation of the amnion-competent cellular blastoderm and the *zen* expression domain (serosa anlage) generated a uniform extraembryonic primordium with early serosal features. This simple model provides an evolutionary framework for understanding gene functions in the amnioserosa. In closing this chapter, we discuss a few examples that may serve to illustrate this point.

After stage 8 (when *zen* is shut off), the maintenance of the amnioserosa depends on genes of the *u-shaped* group (ush-group), such as *u-shaped* (*ush*), *dorsocross* (*doc*), *hindsight* (*hnt*), *tail-up* (*tup*) or *serpent* (*srp*).[41-43] These genes are unrelated but share similar germband-retraction and dorsal-closure phenotypes due to the precocious disintegration of the amnioserosa. All of them are co-expressed with *zen* (see Berkeley *Drosophila* Genome Project at http://www.fruitfly.org/cgi-bin/ex/insitu.pl). However, none of them is required for the specification of the amnioserosa. In *Anopheles*[24] and *Megaselia* (our unpublished data), *doc*, *tup* and *hnt* are expressed in the amnion. Thus, it is possible that the maintenance functions of ush-group genes in the amnioserosa relate to functions of their homologs in the amnion of less derived dipterans.

Two of these genes, *doc* and *hnt*, activate *Krüppel* (*Kr*) in the late amnioserosa.[42,43] In *Megaselia*, the putative ortholog of *Kr* (*Mab-Kr*) is expressed in the early serosa and later in the amnion. In the early serosa, *Mab-Kr* transcripts appear with gastrulation and persist at least until the formation of the amnioserosal fold. In the amnion, expression starts with stomodeum formation (like in *Drosophila*), i.e., after the completion of the serosa and persists at least until germband retraction (Fig. 5). The temporal correlation suggests that *Kr* expression in the amnioserosa is homologous to the amniotic expression in *Megaselia*. Consistent with this hypothesis, *Mab-zen* RNAi embryos lack the early (serosal),[18] but not the late (amniotic) phase of extraembryonic *Mab-Kr* expression.[28] Conversely, serosal *Kr* expression in *Megaselia* (and *Episyrphus*, S.L. unpublished observation) has no equivalent in *Drosophila* and may have been lost in this lineage because postgastrular serosa development is suppressed.

Another example, *pannier* (*pnr*), encodes a GATA transcription factor, which is expressed in the early amnioserosa and in the dorsal epidermis.[44-46] However, in the amnioserosa, the activity of *pnr* appears to be blocked and both the *pnr* transcript and the *pnr* protein disappear during stage 9.[47,48] In *Megaselia* and *Tribolium, pnr* is expressed in the early amnion and adjacent embryonic tissue.[18,23] In *Megaselia*, amniotic *pnr* transcript (protein data are not available) lasts until the early expansion phase of the serosa, which is roughly comparable to the stage when *pnr* is down-regulated in *Drosophila*. Thus, it is possible that the ancestral extraembryonic function of *pnr* was restricted to the early phase of amnion development. According to our model, such a function should have been lost with the conflation of the boundaries of amnion-competent and serosal blastoderm. Hence, we propose that *pnr* expression expanded into the amnioserosa without consequences because another factor in this tissue repressed its activity.

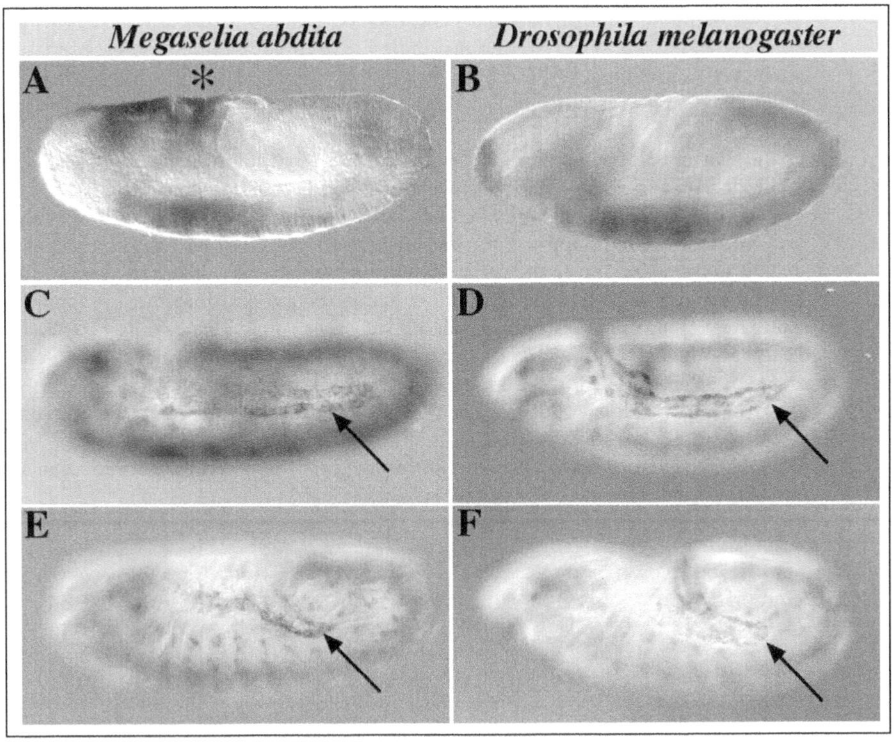

Figure 5. Expression of *Mab-Kr/Kr* transcripts in *Megaselia abdita* and *Drosophila melanogaster*. Matching consecutive stages of *Megaselia* (left) and *Drosophila* (right) are shown during gastrulation (A,B), shortly before germband retraction (C,D) and during germband retraction (E,F). Early serosal (asterisk) and amniotic/amnioserosal *Kr* expression domains (arrows) are marked. Anterior is left and dorsal up.

As a final example of how our hypothesis for the origin of the amnioserosa can provide context for the function of genes in the amnioserosa, we turn to the homeobox gene *C15*, which is activated in the amnioserosa anlage and a narrow strip of adjacent dorsal ectoderm by direct *cis*-regulatory input of *dpp* effectors and Zen.[49] C15 protein persists throughout the lifetime of the amnioserosa,[43] but *C15* null mutations do not interfere with the formation of viable larvae.[50] Thus, *C15* might affect amnioserosa development in minor ways. In *Megaselia, C15* is predominantly expressed in the early amnion and in the dorsal epidermis.[18] Transcript expression in the amnion fades before the expansion of the serosa, which is consistent with a developmental role of this gene in the early amnion. As in the case of *pnr*, the early amniotic function may have been lost in the *Drosophila* lineage as the boundary between amnion-competence and *zen* expression dissolved.

Conclusion

Changes in the expression of *zen* accompanied major reorganizations in extraembryonic development both at the morphological and the regulatory level. At this point, however, the data matrix for any molecular feature of extraembryonic development in insects remains very patchy. With this limitation in mind our tentative conclusions from the above discussion are as follows:

1. The ancestral extraembryonic role of *zen* in pterygote insects was in the mature serosa and in parts of the mature amnion, mediating katatrepsis or a related blastokinetic movement. Holometabola abandoned katatrepsis but retained the functions of *zen* in controlling

rupture of the amniotic cavity and in generating a continuous serosa-amnion epithelium. The loss of amniotic *zen* expression may underlie the suppression of this process in some holometabolous lineages.
2. Zen-dependent serosa specification evolved before or during the early radiation of Holometabola.
3. In the course of dipteran evolution, the ratio of extraembryonic to embryonic blastoderm was reduced and the ratio of *zen*-positive to *zen*-negative amnion-competent blastoderm was increased. The underlying mechanisms involved posterior derepression of *zen* and enhanced transport of Dpp towards the dorsal midline.
4. In schizophoran flies, postgastrular *zen* expression was lost. This change in *zen* expression may have triggered the origin of the amnioserosa in the schizophoran lineage by suppressing postgastrular serosa development, while allowing the completion of a dorsal amnion.
5. After the origin of the amnioserosa, extraembryonic development became entirely dependent on *zen*. This change in the function of *zen* may have evolved in concert with a novel broad dorsal expression domain of *zen* in the early blastoderm and bistable Dpp signaling activity in the cellular blastoderm.
6. The requirement for *zen* activity in the optic field evolved with the loss of *zen*-independent specification of amnion-competent blastoderm.

Acknowledgements

We thank Chip Ferguson for discussions and Kristen A. Panfilio, an anonymous reviewer and Kacy L. Gordon for useful comments on the manuscript.

References

1. Hughes CL, Kaufman TC. Hox genes and the evolution of the arthropod body plan. Evol Dev 2002; 4:459-499.
2. Mann RS, Chan SK. Extra specificity from extradenticle: the partnership between HOX and PBX/EXD homeodomain proteins. Trends Genet 1996; 12:258-62.
3. Moens CB, Selleri L. Hox cofactors in vertebrate development. Dev Biol 2006; 291:193-206.
4. Joshi R, Passner JM, Rohs R et al. Functional specificity of a Hox protein mediated by the recognition of minor groove structure. Cell 2007; 131:530-43.
5. Hughes CL, Liu PZ, Kaufman TC. Expression patterns of the rogue Hox genes Hox3/zen and fushi tarazu. Evol Dev 2004; 6:393-401.
6. Panfilio KA, Akam M. A comparison of Hox3 and Zen protein coding sequences in taxa that span the Hox3/zen divergence. Dev Genes Evol 2007; 217:323-9.
7. Rushlow C, Doyle H, Hoey T et al. Molecular characterization of the zerknüllt region of the Antennapedia gene complex in Drosophila. Genes Dev 1987; 1:1268-1279.
8. Panfilio KA. Extraembryonic development in insects and the acrobatics of blastokinesis. Dev Biol 2008; 313:471-91.
9. Panfilio KA, Liu PZ, Akam M et al. Oncopeltus fasciatus zen is essential for serosal tissue function in katatrepsis. Dev Biol 2006; 292:226-43.
10. Panfilio KA. Late extraembryonic morphogenesis and its zen RNAi-induced failure in the milkweed bug Oncopeltus fasciatus. Dev Biol 2009; 333:297-311.
11. Machida R, Ando H. Evolutionary changes in developmental potentials of the embryo proper and embryonic membranes along with the derivative structures in Atelocerata, with special reference to hexapoda (arthropoda). Proc Arthropod Embryol Soc Jpn 1998; 33:1-12.
12. Sander K. In Insect Development. Lawrence PA, ed. Oxford: Blackwell Scientific Publications, 1976;35-52.
13. Anderson DT. In Developmental Systems: insects. Counce SJ, Waddington CH, eds. London: Academic, 1972;165-242.
14. Davis CWC. A comparative study of larval embryogenesis in the mosquito Culex fatigans Wiedemann (Diptera: Culicidae) and the sheep-fly Lucilia sericata Meigen (Diptera: Calliphoridae). Aust J Zool 1967; 15:547-579.
15. Kahle W. Die Paedogenesis der Cecidomyiden. Zoologica, Stuttgart 1908; 21:1-80 (Plates I-VI).
16. Gambrell FL. The embryology of the blackfly Simulium pinctipes Hagen. Ann Entomol Soc Am 1933; 26:641-671.
17. Butt FH. Embryology of Sciara. Ann Entomol Soc Am 1934; 27:565-579.

18. Rafiqi AM, Lemke SJ, Ferguson S et al. Evolutionary origin of the amnioserosa in cyclorrhaphan flies correlates with spatial and temporal expression changes of zen. Proc Natl Acad Sci USA 2008; 105:234-239.
19. Campos-Ortega JA, Hartenstein V. The embryonic development of Drosophila melanogaster (Springer-Verlag, Berlin, Heidelberg, New York, 1997).
20. Reed BH, Wilk R, Schöck F et al. Integrin-dependent apposition of Drosophila extraembryonic membranes promotes morphogenesis and prevents anoikis. Curr Biol 2004; 14:372-380.
21. Falciani F, Hausdorf B, Schröder R et al. Class 3 Hox genes in insects and the origin of zen. Proc Natl Acad Sci USA 1996; 93:8479-8484.
22. Dearden P, Grbic M, Falciani F et al. Maternal expression and early zygotic regulation of the Hox3/zen gene in the grasshopper Schistocerca gregaria. Evol Dev 2000; 2:261-270.
23. van der Zee M, Berns N, Roth S. Distinct functions of the Tribolium zerknüllt genes in serosa specification and dorsal closure. Curr Biol 2005; 15:624-636.
24. Goltsev Y, Fuse N, Frasch M et al. Evolution of the dorsal-ventral patterning network in the mosquito, Anopheles gambiae. Development 2007;2415-2424.
25. Stauber M, Prell A, Schmidt-Ott U. A single Hox3 gene with composite bicoid and zerknüllt expression characteristics in non-Cyclorrhaphan flies. Proc Natl Acad Sci USA 2002; 99:274-279.
26. Brown S, Fellers J, Shippy T et al. A strategy for mapping bicoid on the phylogenetic tree. Curr Biol 2001; 11:R43-R44.
27. Pultz, MA Diederich RJ, Cribbs DL et al. The proboscipedia locus of the antennapedia complex: a molecular and genetic analysis. Genes Dev 1988; 2:901-920.
28. Rafiqi AM. Morphological transitions and the genetic basis of the evolution of extraembryonic tissues in flies. Ph.D. Thesis, Wageningen University (NL) 2008.
29. Stauber M, Lemke S, Schmidt-Ott U. Expression and regulation of caudal in the lower cyclorrhaphan fly Megaselia. Dev Genes Evol 2008; 218:81-7.
30. Newfeld SJ, Wisotzkey RG, Kumar S. Molecular evolution of a developmental pathway: phylogenetic analyses of transforming growth factor-beta family ligands, receptors and Smad signal transducers. Genetics 1999; 152:783-95.
31. O'Connor MB, Umulis D, Othmer HG et al. Shaping BMP morphogen gradients in the Drosophila embryo and pupal wing. Development 2006; 133:183-93.
32. Ferguson EL, Anderson KV. Decapentaplegic acts as a morphogen to organize dorsal-ventral pattern in the Drosophila embryo. Cell 1992; 71:451-61.
33. Dorfman R, Shilo BZ. Biphasic activation of the BMP pathway patterns the Drosophila embryonic dorsal region. Development 2001; 128:965-972.
34. Tanimoto H, Itoh S, ten Dijke P et al. Hedgehog creates a gradient of DPP activity in Drosophila wing imaginal discs. Mol Cell 2000; 5:59-71.
35. Persson U, Izumi H, Souchelnytskyi S et al. The L45 loop in type I receptors for TGF-beta family members is a critical determinant in specifying Smad isoform activation. FEBS Lett 1998; 434:83-7.
36. Wang Y-C, Ferguson EL. Spatial bistability of Dpp-receptor interactions during Drosophila dorsal-ventral patterning. Nature 2005; 434:229-234.
37. van der Zee M, Stockhammer O, von Levetzow C et al. Sog/Chordin is required for ventral-to-dorsal Dpp/BMP transport and head formation in a short germ insect. Proc Natl Acad Sci USA 2006; 103:16307-16312.
38. Schoppmeier M, Schröder R. Maternal torso signaling controls body axis elongation in a short germ insect. Curr Biol 2005; 15:2131-6.
39. Angelini DR, Kaufman TC. Functional analyses in the milkweed bug (Hemiptera) support a role for Wnt signaling in body segmentation but not appendage development. Dev Biol 2005; 283:409-23.
40. Chang T, Mazotta J, Dumstrei K et al. Dpp and Hh signaling in the Drosophila embryonic eye field. Development 2001; 128:4691-4704.
41. Frank LH, Rushlow C. A group of genes required for maintenance of the amnioserosa tissue in Drosophila. Development 1996; 122:1343-1352.
42. Yip MLR, Lamka ML, Lipshitz HD. Control of germ-band retraction in Drosophila by the zinc-finger protein Hindsight. Development 1997; 124:2129-2141.
43. Reim I, Lee H-H, Frasch M. The T-box-encoding dorsocross genes function in amnioserosa development and the patterning of the dorsolateral germ band downstream of Dpp. Development 2003; 130:3187-3204.
44. Winick J, Abel T, Leonard MW et al. A GATA family transcription factor is expressed along the embryonic dorsoventral axis in Drosophila melanogaster. Development 1993; 119:1055-1065.
45. Heitzler P, Haenlin M, Ramain P et al. A genetic analysis of pannier, a gene necessary for viability of dorsal tissues and bristle positioning in Drosophila. Genetics 1996; 143:1271-1286.

46. Ashe HL, Mannervik M, Levine M. Dpp signaling thresholds in the dorsal ectoderm of the Drosophila embryo. Development 2000; 127:3305-12.
47. Haenlin M, Cubadda Y, Blondeau F et al. Transcriptional activity of pannier is regulated negatively by heterodimerization of the GATA DNA-binding domain with a cofactor encoded by the u-shaped gene of Drosophila. Genes Dev 1997; 11:3096-108.
48. Herranz H, Morata G. The functions of pannier during Drosophila embryogenesis. Development 2001; 128:4837-46.
49. Lin MC, Park J, Kirov N et al. Threshold response of C15 to the Dpp gradient in Drosophila is established by the cumulative effect of Smad and Zen activators and negative cues. Development 2006; 133:4805-13.
50. Campbell G. Regulation of gene expression in the distal region of the Drosophila leg by the Hox11 homolog, C15. Dev Biol 2005; 278:607-18.
51. Grimaldi D, Engel MS. Evolution of insects. Cambridge: Cambridge University Press, 2005.

Chapter 11

Hox Genes and Brain Development in Drosophila

Heinrich Reichert* and Bruno Bello

Abstract

Hox genes are prominently expressed in the developing brain and ventral ganglia of *Drosophila*. In the embryonic brain, the Hox genes *labial* and *Deformed* are essential for the establishment of regionalized neuronal identity; in their absence cells are generated in the brain but fail to acquire appropriate neuronal features. Genetic analyses reveal that Hox proteins are largely equivalent in their action in embryonic brain development and that their expression is under the control of cross-regulatory interactions among Hox genes that are similar to those found in embryogenesis of trunk segments. Hox genes have a different role in postembryonic brain development. During the larval phase of CNS development, reactivation of specific Hox genes terminates neural proliferation by induction of apoptotic cell death in neural stem cell-like progenitors called neuroblasts. This reactivation process is tightly controlled by epigenetic mechanisms requiring the Polycomb group of genes. Many features of Hox gene action in *Drosophila* brain development are evolutionarily conserved and are manifest in brain development of vertebrates.

Introduction

Hox genes encode a network of conserved transcription factors that are involved in specifying regional identity along the anteroposterior embryonic body axis of animals as diverse as insects and vertebrates.[1-3] Hox genes were first discovered in *Drosophila* where the genes are arranged along the chromosome in two gene clusters known as the Antennapedia (ANT-C) and Bithorax (BX-C) Complexes. In this genetic model system, as in many other bilaterians, there is a remarkable correlation between the relative position of the Hox genes in the clusters and their spatial and temporal expression patterns in the embryo in that genes located towards the 3' end are expressed more anteriorly and earlier than genes towards the 5' end; this is referred to as spatial and temporal colinearity.[4]

In *Drosophila*, as for most bilaterian animals, Hox genes are prominently expressed in the embryonic central nervous system (CNS), which in the fly is comprised of the brain and ventral ganglia. A number of recent genetic analyses have addressed the functional roles of Hox genes in the development of the fly CNS. In this review, we will focus on the role of the Hox genes in the developing *Drosophila* brain. For an excellent recent review of the action of Hox genes (and other patterning genes) in the development of the ventral ganglia of *Drosophila* we refer the reader to reference 5.

*Corresponding Author: Heinrich Reichert—Biozentrum, University of Basel, Klingelbergstrasse 50, CH-4056 Basel, Switzerland. Email: heinrich.reichert@unibas.ch

Hox Genes: Studies from the 20th to the 21st Century, edited by Jean S. Deutsch.

Expression and Function of Hox Genes in Embryonic Brain Development

The *Drosophila* brain is composed of an anterior supraesophageal ganglion and a posterior subesophageal ganglion. The supraesophageal ganglion is subdivided into the protocerebrum, the deutocerebrum and the tritocerebrum; the subesophageal ganglion is subdivided into the mandibular, maxillary and labial neuromeres.[6,7] In the embryonic brain of *Drosophila,* all five Hox genes of the ANT-C *(labial, proboscipedia, Deformed, Sex combs reduced, Antennapedia)* are expressed in discrete domains of specific neuromeres and their anterior expression boundaries often coincide with neuromere compartment boundaries (Hirth et al, 1998; Urbach and Technau, 2003a; Sprecher et al, 2007).[8-10] In contrast to the embryonic epidermal structures of *Drosophila,* the anteroposterior arrangement of the homeotic genes in the fly CNS does not strictly fulfill the criterion of spatial colinearity.[8,11] The expression domains of the two 3' most Hox genes of the ANT-C are inverted in that the anterior expression boundary of *labial* is posterior to that of *proboscipedia.* A summary of the expression patterns of all Hox genes in the embryonic brain and ventral ganglia is shown in Figure 1.

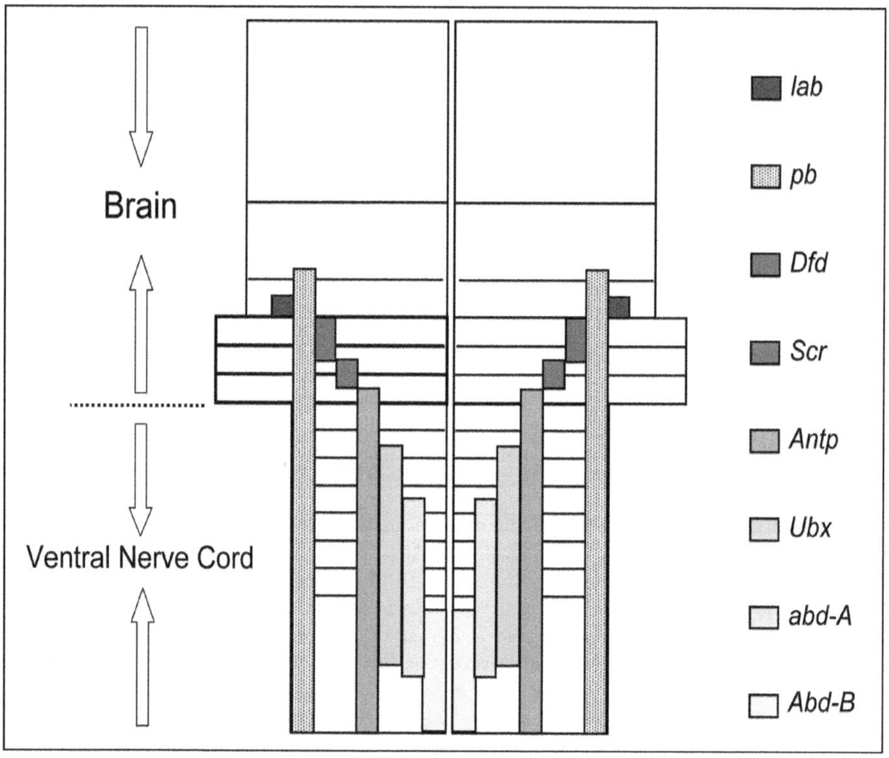

Figure 1. Expression of Hox genes in embryonic brain development. Schematic diagram of the anteroposterior order of gene expression in the brain and ventral nerve cord of a stage 14 *Drosophila* embryo. Expression domains of the genes *lab, pb, Dfd, Scr, Antp, Ubx, abd-A* and *Abd-B* are indicated by shaded bars. Segmental borders in the brain and anterior ventral ganglia are indicated by horizontal lines. In contrast to the other Hox genes, *pb* is expressed only in small segmentally repeated groups of neural cells; this difference is indicated by dotted shading. Anterior is to the top.

Loss-of-function mutant analyses have been carried out for all five genes expressed in the embryonic brain and in the case of *lab* (*labial*) and *Dfd* (*Deformed*) dramatic mutant phenotypes have been uncovered.[8] Thus, in *lab* null mutants, axonal projection defects occur in the region of the tritocerebrum in which *lab* is normally expressed in the wild-type brain. In the mutant, longitudinal pathways connecting supraesophageal and subesophageal ganglia as well as the projections in the tritocerebral commissure are absent or reduced. Moreover, the frontal connectives no longer project into the tritocerebral neuromere but rather grow ectopically into the more anterior brain neuromeres. Interestingly, the brain defects are not due to a deletion in the tritocerebral neuromere; neuronal progenitors are present and give rise to postmitotic progeny in the mutant domain. These postmitotic cells, however, do not form axonal and dendritic extensions and are not contacted by axons from other parts of the brain. Indeed, the *lab* mutant cells do not acquire a neuronal fate, as revealed by the absence of neuronal markers, but rather remain undifferentiated (Fig. 2). This

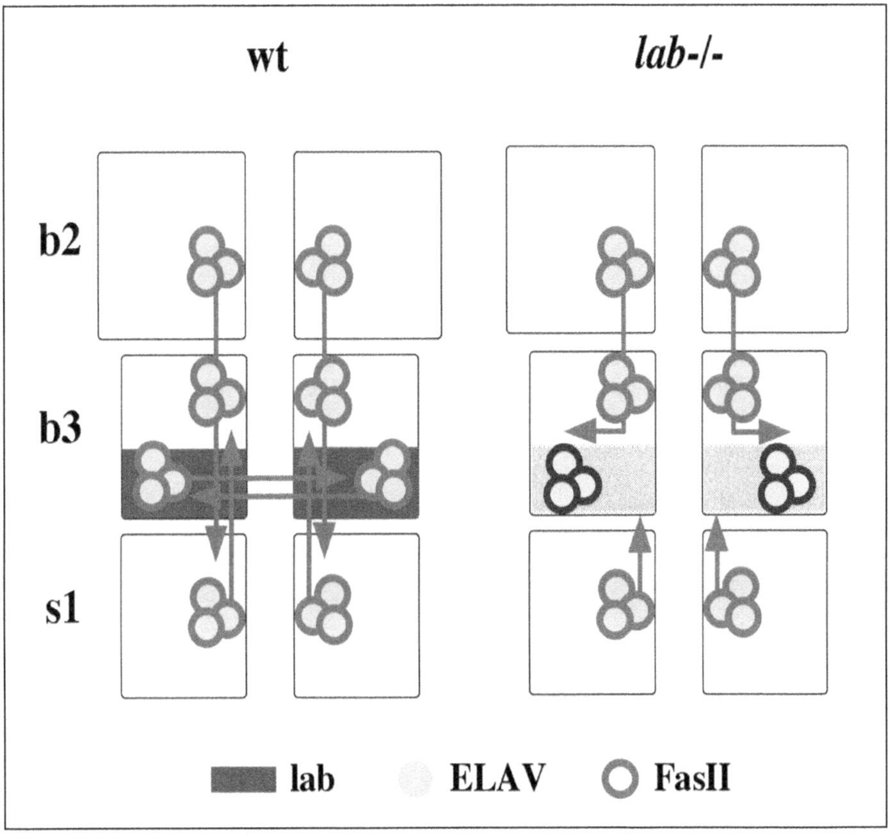

Figure 2. Function of the Hox gene *lab* in embryonic brain development. Simplified scheme of the deutocerebral (b2), tritocerebral (b3) and mandibular (s1) neuromeres in the embryonic *Drosophila* brain. In the wild-type (wt) cells in the posterior tritocerebrum express *lab* (dark shading) and also express the neuron-specific marker ELAV and the cell adhesion molecule fasciclin II (FasII). Axons from other parts of the brain (arrows) project to the *lab* domain. In the *lab* null mutant *(lab–/–)*, cells in the mutant domain (light shading) are present but do not extend axons and fail to express the neuron-specific marker ELAV and the cell adhesion molecule fasciclin II, indicating a total loss of neuronal identity. Axons from other parts of the brain avoid the mutant domain. Anterior is to the top.

suggests that the pattern of proliferation in the tritocerebrum is initiated correctly in the absence of the *lab* gene product, but that the cells that normally express *lab* do not become correctly specified in the mutant leading to severe defects in axonogenesis. Comparable defects are seen in the *Dfd* mutant in the corresponding mandibular domain, where the wild-type expression of the gene is located. Thus, the appropriate expression of the homeotic genes *lab* and *Dfd* is essential for the establishment of regionalized neuronal identity in the embryonic brain of *Drosophila*.

Remarkably, there is a significant degree of functional equivalence of the Hox gene products in the specification of neuronal identity during embryonic brain development. Thus, when placed under the control of appropriate CNS-specific regulatory elements, all Hox proteins except AbdB are able to efficiently replace Lab protein in the specification of the tritocerebral neuromere.[12] While there is some correlation between the replacement efficiency of the Hox proteins and the chromosomal arrangement of their encoding loci, these findings indicate that despite considerably diverged sequences, most Hox proteins are functionally comparable in their ability to replace Lab in the specification of neuronal identity. This suggests that in embryonic brain development, differences in Hox gene action might rely mainly on *cis*-acting regulatory elements and not on Hox protein specificity.

Genetic Interactions between Hox Genes in Embryonic Brain Development

In studies of morphogenesis in *Drosophila,* Hox genes interact both genetically and molecularly. The term "posterior prevalence" (also referred to as "phenotypic suppression") has been proposed to describe the cross-regulation of these genes and the phenotypic consequences of their expression. Posterior prevalence describes the down regulation of a more anteriorly expressed Hox gene by those that are more posteriorly expressed.[13-15] Furthermore, Hox transcription factors in *Drosophila* often bind to DNA as a heterodimer with another homeodomain protein encoded by the *extradenticle* (*exd*) gene. When the Exd cofactor binds together with Hox proteins, it increases their DNA binding specificity and affinity and also modifies their transcriptional regulatory properties.[16,17] A further homeodomain protein that may interact with Exd/Hox heterodimers is encoded by the *homothorax* (*hth*) gene. Hth and Exd proteins directly interact with each other and the nuclear localization of Exd depends on this interaction.[18,19]

During embryonic brain development expression of the *lab* gene in the tritocerebrum is subject to posterior dominance cross-regulatory interactions. Early ectopic expression of *Ultrabithorax* (*Ubx*) or *abdominal-A* (*abd-A*) represses *lab* expression in the embryonic CNS in a timing-dependent manner.[20] Genetic interactions between *lab* and *exd/hth* occur in the developing tritocerebrum; mutations of *exd* or *hth* result in suppression of *lab* expression in this brain neuromere.[21] Moreover, specification of the tritocerebral neuronal identity appears to require balanced levels of Hox proteins and Hth and nuclear-targeted n-Exd cofactors. Thus, misexpression of posterior Hox genes (as opposed to replacement of Lab by posterior Hox proteins; see above) results in a *lab* loss-of function phenotype in the developing tritocerebrum correlated with a lack of Lab protein expression in the tritocerebrum.[22] This lack of Lab protein is due to repression of transcription of the *lab* gene in the embryonic brain. In this respect this phenomenon differs from the canonical form of posterior prevalence which acts at the postranslational and not at the transcriptional level. Interestingly, the repressive activity that underlies this posterior prevalence effect of Hox genes on *lab* expression can be abolished by the concomitant targeted misexpression of the cofactors Hth and n-Exd. These findings in embryonic brain development have many features of the Hox gene interacting networks that have been studied in detail in embryogenesis of trunk segments and, thus, provide support for a general model for the regulation of Hox gene activity that involves Hox genes, cofactors and Hox target elements.

Contrasting with the demonstrated interactions between *lab* and other different Hox genes in embryonic brain development, are findings that indicate a lack of interaction between the Hox gene *lab* and the homeobox-containing columnar gene *vnd* (*ventral nervous system defective*). The *vnd* gene is one of three columnar genes which act in dorsoventral patterning of the CNS by dividing the embryonic neuroectoderm along its dorsoventral axis into adjacent longitudinal columns.[23] The *lab* and *vnd* genes show overlapping expression in the tritocerebral neuroblasts (primary progenitors) and subsequently in neural cells of the tritocerebrum and both *lab* and *vnd* mutant brain phenotypes result in comparable axonal patterning defects.[24-26] Nevertheless, the genetic activity of these two patterning genes appears to be mutually independent. Thus, *vnd* may be required for the specification of neural lineages within the developing tritocerebral neuromere during early stages of embryonic neurogenesis, whereas *lab* may be independently required for the establishment of regionalized neuronal identity within the same territory during later stages.[26] This implies that the activity of the dorsoventral patterning gene *vnd* is integrated into pattern formation along the anteroposterior neuraxis by ensuring proper formation and development of neural lineages that subsequently become further specified by the activity of the Hox gene *lab*.

Hox Genes in Postembryonic Brain Development

During embryogenesis, the neurons of the larval brain and ventral ganglia are generated by a set of stem cell-like neuroblasts. Following a short period of quiescence, most of these neuroblasts resume neural proliferation during larval development and, during this second phase of neurogenesis, generate the bulk of the adult CNS postembryonically.[27,28] Although Hox genes play important roles in embryonic neurogenesis, we still know very little about Hox gene expression and action in postembryonic CNS development. However, recently some insight into the role of Hox genes in temporal regulation of postembryonic neural proliferation has been obtained.

It has long been proposed that regulation of cell proliferation by Hox genes is an important factor in shaping segment-specific morphologies in animals as diverse as insects and vertebrates.[29] Indeed, during neurogenesis, temporal regulation of proliferation plays an important role in ensuring that the appropriate number of neural progeny are generated by each neuroblast and, hence, assigned to each specialized region of the CNS.[30] Although the genetic pathways have yet to be worked out in detail, it is clear that region-specific molecular inputs, such as those provided by Hox genes, are likely to be important in this temporal regulation process.

In accordance with this assumption, recent findings show that reactivation of specific Hox genes is involved in terminating neuroblast proliferation during postembryonic development.[31] These findings indicate that specific posterior neuroblasts in the wild-type undergo programmed cell death during late larval stages (Fig. 3). This is because a pulse of synthesis of the Hox protein AbdA is required in these dividing neuroblasts to specify the time at which apoptosis occurs, thereby determining the final number of progeny that each neuroblast generates. Accordingly, when these neuroblasts are made deficient for the three proapoptotic genes, *head involution defective*, *grim* and *reaper*, they persist throughout larval life and generate many more neuronal progeny than normal. Since Hox proteins other than AbdA also have this intrinsic ability to trigger neuroblast-specific death, this strategy for regulation of neural number during postembryonic development is likely to be used in other regions of the CNS including the brain.

Given this direct link between neuromere-specific Hox gene expression and the induction of apoptosis in neuroblasts, tight regulation of Hox gene expression during CNS development is likely to be essential. How might this regulation be mediated? In many cases, tissue-specific gene expression is regulated through the stable maintenance of an active or repressed state by epigenetic mechanisms. Part of this epigenetic regulation involves the Polycomb group (PcG) of genes, which regulate gene expression by the stable silencing of target genes through chromatin modifications.[32,33] PcG genes were originally discovered in *Drosophila* as repressors of the Hox genes and deregulation of Hox genes' expression is one of the hallmarks of PcG mutant phenotypes in both *Drosophila* and vertebrates. Based on their prominent roles in maintaining appropriate spatial patterns of

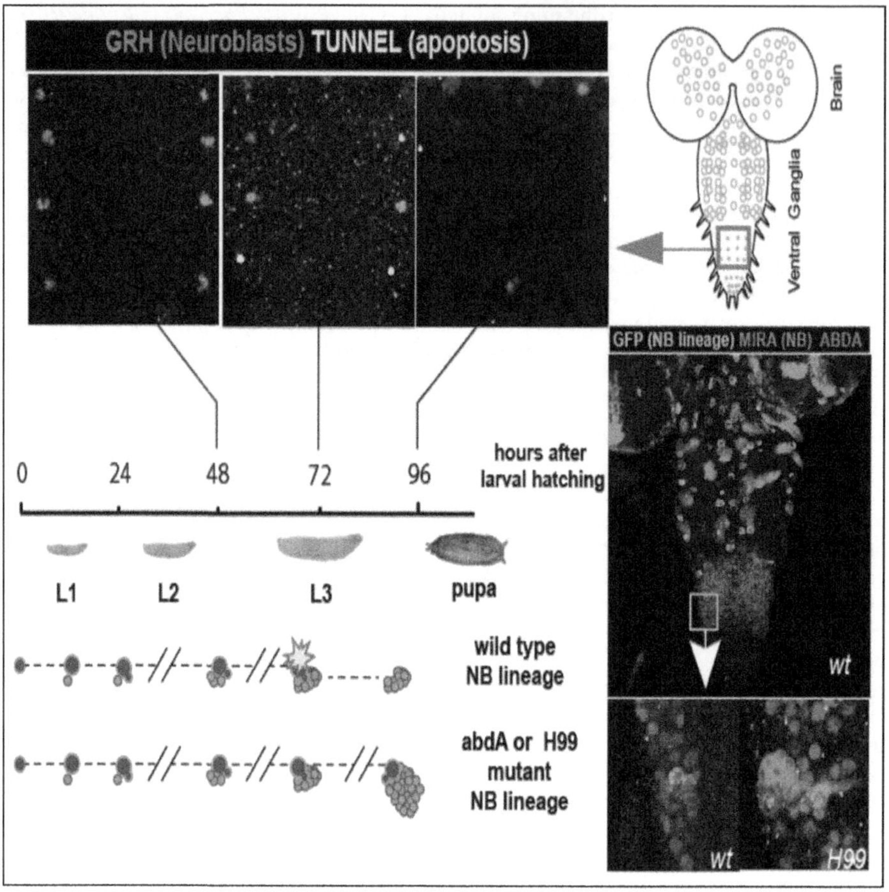

Figure 3. Reactivation of the Hox gene *abd-A* is involved in terminating neuroblast proliferation during postembryonic development in *Drosophila*. In the wild-type, apoptosis of specific neuroblasts occurs during the larval development; apopototic cell death in larval CNS neuroblasts is assayed by TUNNEL staining (top left) and schematic of neuroblasts in larval CNS (top right). Blocking apoptosis genetically or clonal mutation of the *abd-A* gene results in neuroblast survival; schematic summary diagram (bottom left) and genetically labelled neuroblast lineages (bottom right).

Hox genes' expression, it is likely that PcG genes might also play a role in controlling neuronal proliferation in brain development.

This notion is supported by recent findings that show that postembryonic neuroblasts fail to proliferate normally and are eliminated by apoptosis in the absence of PcG genes.[34] This proliferation defect can be rescued by blocking apoptosis in these neuroblasts indicating that PcG genes are required to prevent neuroblast death. Importantly, aberrant ectopic expression of posterior Hox genes (which in wild-type leads to neuroblast death; see above) occurs in PcG mutant neuroblasts as well as in PcG mutant neuroblasts rescued by apoptosis-block. These findings indicate that loss of PcG genes leads to aberrant derepression of Hox genes expression in postembryonic neuroblasts, resulting in neuroblast death and termination of proliferation in mutant lineages (Fig. 4). Taken together, these findings imply that repression of aberrant

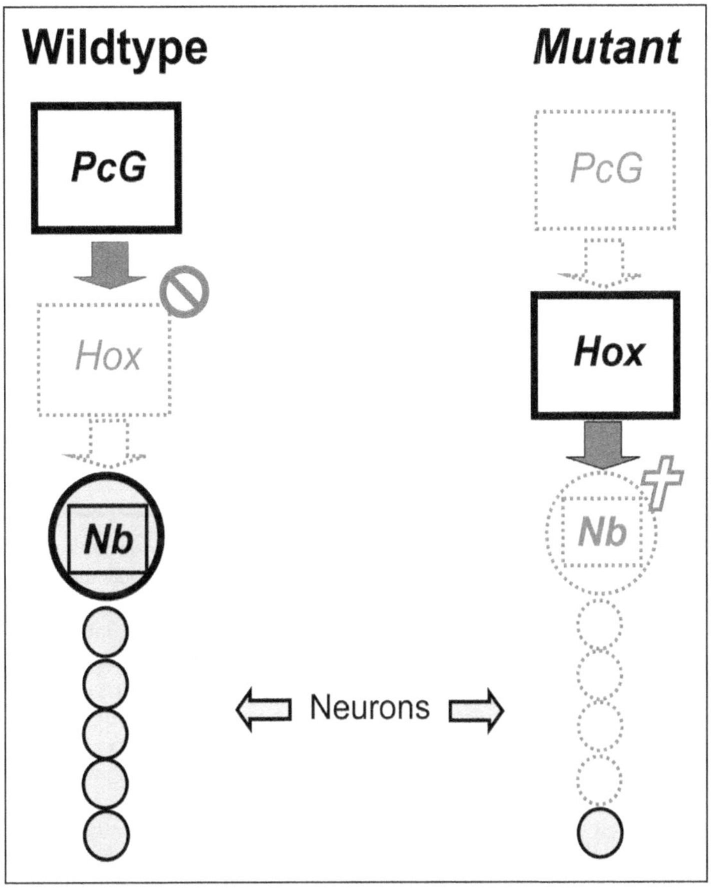

Figure 4. PcG genes repress aberrant Hox gene activation in postembryonic brain development of *Drosophila*. Simplified summary scheme. In the wild-type (left) PcG (Polycomb group) genes prevent the ectopic expression of Hox genes in postembryonic neuroblasts (Nb) and, hence, allow neuroblasts to survive and proliferate normally to produce neurons. In PcG mutants (right) loss of PcG genes leads to an aberrant derepression of Hox gene activity in postembryonic neuroblasts, resulting in neuroblast death and termination of neuron proliferation in these lineages.

reactivation of Hox gene expression is a crucial role of PcG genes in postembryonic brain and CNS development.

Evolutionary Conservation of Hox Gene Action in Brain Development

Hox gene expression in the developing CNS is a shared feature of a wide range of bilaterian animals, including protostomes such as insects or annelids and deuterostomes, such as hemichordates or vertebrates.[8,35-37] Remarkably, throughout the Bilateria, Hox gene orthologs are expressed in a similar anteroposterior order in the developing CNS. For example, in the mouse, Hox genes are expressed in the developing hindbrain and spinal cord and their relative anteroposterior order of expression in the developing CNS is very similar to their arrangement in the *Drosophila* CNS, including the inverted order of the *lab* and *pb* orthologs, *Hoxb1* and

Hoxb2.[38] As more expression data from different protostome and deuterostome species becomes available, the ordered expression of Hox genes along the anteroposterior axis of the developing nervous system is likely to consolidate as a common feature of bilaterian animals.

In *Drosophila*, mutational inactivation of either of the homeotic genes *lab* or *Dfd* causes severe axonal patterning defects in the embryonic brain (see ref. 8). Mutational inactivation of the murine *lab* orthologs, *Hoxa1* and *Hoxb1*, which are expressed in overlapping domains in the developing hindbrain, also causes defects in embryonic brain development. Functional inactivation of *Hoxa1* results in segmentation defects leading to a reduced size of rhombomeres 4 and 5 and defects in motor neuron axonal projections, but the normal identity of rhombomere 4 is not altered.[39] In contrast, loss of *Hoxb1* function has no influence on the size of rhombomere 4 but causes a partial transformation into a rhombomere 2 identity.[40] The *Hoxa1*; *Hoxb1* double mutant results in a territory of unknown identity and reduced size between rhombomeres 3 and 5, suggesting a synergistic action of the two genes in rhombomere 4 specification.[39] Thus, the concerted activity of *Hoxa1* and *Hoxb1* appears to have a similar role in the specification of the regionalized neuronal identity as does their ortholog *lab* in the CNS of *Drosophila*. This suggests a functional conservation of Hox gene action, in addition to a similar mode of expression, during brain development of bilaterian animals.

Conclusion

A wealth of experimental data indicates that Hox genes play important roles in the development of the Drosophila brain. Remarkably, these roles are distinctly different during embryonic and postembryonic development. Hox genes are involved in establishing regionalized identity of neurons in specific neuromeres of the embryonic brain; in contrast they control the termination of neuronal proliferation by inducing apoptotic cell death of neuroblasts in postembryonic brain development. In the developing brain, Hox gene action is itself tightly controlled by complex regulatory processes that involve specific cofactors, Hox gene cross-regulatory interactions, and epigenetic silencing through Polycomb group genes. Comparative studies indicate that many aspects of Hox gene expression and function in brain development are conserved and are likely to be general features of brain development in bilaterian animals. Given the unexpected diversity and generality of Hox gene action in brain development revealed by recent investigations, it seems safe to predict that future studies will continue to uncover novel roles for these central developmental regulatory genes in the development of the brain.

Acknowledgement

This work was supported by the Swiss NSF.

References

1. McGinnis W, Krumlauf R. Homeobox genes and axial patterning. Cell 1992; 68(2):283-302.
2. Manak JR, Scott MP. A class act: conservation of homeodomain protein functions. Dev Suppl 1994:61-77.
3. Carroll SB. Homeotic genes and the evolution of arthropods and chordates. Nature 1995; 376(6540):479-85.
4. Mann RS. Why are Hox genes clustered? Bioessays 1997; 19(8):661-4.
5. Technau GM, Berger C, Urbach R. Generation of cell diversity and segmental pattern in the embryonic central nervous system of Drosophila. Dev Dyn 2006; 235(4):861-9.
6. Younossi-Hartenstein A, Nassif C, Green P et al. Early neurogenesis of the Drosophila brain. J Comp Neurol 1996; 370(3):313-29.
7. Reichert H, Boyan G. Building a brain: developmental insights in insects. Trends Neurosci 1997; 20(6):258-64.
8. Hirth F, Hartmann B, Reichert H. Homeotic gene action in embryonic brain development of Drosophila. Development 1998; 125(9):1579-89.
9. Urbach R, Technau GM. Molecular markers for identified neuroblasts in the developing brain of Drosophila. Development 2003a; 130(16):3621-37.
10. Sprecher SG, Reichert H, Hartenstein V. Gene expression patterns in primary neuronal clusters of the Drosophila embryonic brain. Gene Expr Patterns 2007; 7(5):584-95.

11. Kaufman TC, Seeger MA, Olsen G. Molecular and genetic organization of the Antennapedia gene complex of Drosophila melanogaster. Adv Genet 1990; 27:309-62.
12. Hirth F, Loop T, Egger B et al. Functional equivalence of Hox gene products in the specification of the tritocerebrum during embryonic brain development of Drosophila. Development 2001; 128(23):4781-88.
13. Duboule D, Morata G. Colinearity and functional hierarchy among genes of the homeotic complexes. Trends Genet 1994; 10(10):358-64.
14. Graba Y, Aragnol D, Pradel J. Drosophila Hox complex downstream targets and the function of homeotic genes. Bioessays 1997; 19(5):379-88.
15. Mann RS, Morata G. The developmental and molecular biology of genes that subdivide the body of Drosophila. Annu Rev Cell Dev Biol 2000; 16:243-71.
16. Mann RS, Chan SK. Extra specificity from extradenticle: the partnership between HOX and PBX/EXD homeodomain proteins. Trends Genet 1996; 12(7):258-62.
17. Pinsonneault J, Florence B, Vaessin H et al. A model for extradenticle function as a switch that changes HOX proteins from repressors to activators. EMBO J 1997; 16(8):2032-42.
18. Mann RS, Affolter M. Hox proteins meet more partners. Curr Opin Genet Dev 1998; 8(4):423-9.
19. Ryoo HD, Marty T, Casares F et al. Regulation of Hox target genes by a DNA bound Homothorax/Hox/Extradenticle complex. Development 1999; 126(22):5137-48.
20. Miller DF, Rogers BT, Kalkbrenner A et al. Cross-regulation of Hox genes in the Drosophila melanogaster embryo. Mech Dev 2001; 102(1-2):3-16.
21. Nagao T, Endo K, Kawauchi H et al. Patterning defects in the primary axonal scaffolds caused by the mutations of the extradenticle and homothorax genes in the embryonic Drosophila brain. Dev Genes Evol 2000; 210(6):289-99.
22. Sprecher SG, Müller M, Kammermeier L et al. Hox gene cross-regulatory interactions in the embryonic brain of Drosophila. Mech Dev 2004; 121(6):527-36.
23. Skeath JB, Thor S. Genetic control of Drosophila nerve cord development. Curr Opin Neurobiol 2003; 13(1):8-15.
24. Urbach R, Technau GM. Segment polarity and DV patterning gene expression reveals segmental organization of the Drosophila brain. Development 2003b; 130(16):3607-20.
25. Urbach R, Volland D, Seibert J et al. Segment-specific requirements for dorsoventral patterning genes during early brain development in Drosophila. Development 2006; 133(21):4315-30.
26. Sprecher SG, Urbach R, Technau GM et al. The columnar gene vnd is required for tritocerebral neuromere formation during embryonic brain development of Drosophila. Development 2006; 133(21):4331-9.
27. Prokop A, Technau GM. The origin of postembryonic neuroblasts in the ventral nerve cord of Drosophila melanogaster. Development 1991; 111(1):79-88.
28. Truman JW, Bate M. Spatial and temporal patterns of neurogenesis in the central nervous system of Drosophila melanogaster. Dev Biol 1988; 125(1):145-57.
29. Duboule D. Vertebrate Hox genes and proliferation: an alternative pathway to homeosis? Curr Opin Genet Dev 1995; 5(4):525-8.
30. Maurange C, Gould AP. Brainy but not too brainy: starting and stopping neuroblast divisions in Drosophila. Trends Neurosci 2005; 28(1):30-6.
31. Bello BC, Hirth F, Gould AP. A pulse of the Drosophila Hox protein Abdominal-A schedules the end of neural proliferation via neuroblast apoptosis. Neuron 2003 23; 37(2):209-19.
32. Orlando V. Polycomb, epigenomes and control of cell identity. Cell 2003; 112(5):599-606.
33. Ringrose L, Paro R. Polycomb/Trithorax response elements and epigenetic memory of cell identity. Development 2007; 134(2):223-32.
34. Bello B, Holbro N, Reichert H. Polycomb group genes are required for neural stem cell survival in postembryonic neurogenesis of Drosophila. Development 2007; 134(6):1091-9.
35. Kourakis MJ, Master VA, Lokhorst DK et al. Conserved anterior boundaries of Hox gene expression in the central nervous system of the leech Helobdella. Dev Biol 1997; 190(2):284-300.
36. Lowe CJ, Wu M, Salic A et al. Anteroposterior patterning in hemichordates and the origins of the chordate nervous system. Cell 2003; 113(7):853-65.
37. Wilkinson DG, Bhatt S, Cook M et al. Segmental expression of Hox-2 homoeobox-containing genes in the developing mouse hindbrain. Nature 1989; 341(6241):405-9.
38. Hunt P, Krumlauf R. Deciphering the Hox code: clues to patterning branchial regions of the head. Cell 1991; 66(6):1075-8.
39. Gavalas A, Studer M, Lumsden A et al. Hoxa1 and Hoxb1 synergize in patterning the hindbrain, cranial nerves and second pharyngeal arch. Development 1998; 125(6):1123-36.
40. Studer M, Gavalas A, Marshall H et al. Genetic interactions between Hoxa1 and Hoxb1 reveal new roles in regulation of early hindbrain patterning. Development 1998; 125(6):1025-36.

CHAPTER 12

Homeosis and Beyond.
What Is the Function of the Hox Genes?

Jean S. Deutsch*

Abstract

What is the function of the Hox genes? At first glance, it is a curious question. Indeed, the answer seems so obvious that several authors have spoken of 'the Hox function' about some of the Hox genes, namely *Hox3/zen* and *Hox6/ftz* that seem to have lost it during the evolution of Arthropods. What these authors meant is that these genes have lost their '*homeotic*' function. Indeed, 'homeotic' refers to a functional property that is so often associated with the Hox genes. However, the word 'Hox' should not be used to refer to a function, but to a group of genes. The above examples of *Hox3/zen* (see Schmitt-Ott's chapter, this book) and *Hox6/ftz* show that the homeotic function may be not so tightly linked to the Hox genes. Reversely, many genes, not belonging to the Hox group, do present a homeotic function.

In the present chapter, I will first give a definition of the Hox genes. I will then ask what is the 'function' of a gene, examining its various meanings at different levels of biological organization. I will review and revisit the relation between the Hox genes and homeosis. I will suggest that their morphological homeotic function has been secondarily derived during the evolution of the Bilateria.

What Are the Hox Genes?

In *Drosophila melanogaster*, genetic analysis showed that homeotic genes are grouped on the third chromosome in two complexes, the Bithorax-Complex (BX-C)[1] and the Antennapedia-Complex (ANT-C).[2] The discovery of the homeobox[3,4] a common motif in the sequence of genes belonging to both BX-C and ANT-C complexes, showed that these genes are phylogenetically related, supporting Ed Lewis' hypothesis[1] that they are issued from duplications of an ancestral gene. Noteworthy, in addition to the canonical homeotic genes *Ultrabithorax* (*Ubx*) and *Antennapedia* (*Antp*), representatives of the BX-C and ANT-C complexes, respectively, the third gene where the homeobox was initially found is *fushi-tarazu* (*ftz*), a gene that, although located within the ANT-C, has no homeotic function in *Drosophila*. Soon after, the homeobox motif allowed the recovery of related genes in other animals, including vertebrates.[5] The Hox acronym was coined to name those vertebrate genes, then all genes found in a diversity of animals, which are closely related to the first three homeobox-containing *Drosophila* genes.

It soon appeared that a more distantly related homeobox motif is present in a huge variety of genes, even outside the animal kingdom. Thus, homeobox-containing genes comprise numerous families,[6-8] of which the so-called 'Hox' are but a sub-group. I will take here as a definition of Hox genes proper, those among homeobox-containing genes that can be assumed to be orthologs of the 13 'paralogy groups' found in tetrapod vertebrates,[9,10] to which a 14th member, found in

*Jean S. Deutsch—Emeritus Professor, Université P. and M. Curie (Paris 6). CNRS, UMR 7622 Biologie du développement. Email: jean.deutsch@snv.jussieu.fr

Hox Genes: Studies from the 20th to the 21st Century, edited by Jean S. Deutsch.

nontetrapod chordates, has now to be added[11] (see Ferrier's chapter, this book). As a corollary, distinction between the Hox genes proper and other homeobox-containing genes must be based on phylogenetic analyses. Using the homeodomain sequences, it appears that a well-defined monophyletic group of genes, that here I will call the Hox-extended family, includes all the Hox genes, but also others. Among these are the so-called paraHox genes[12] and the *Mox/Btn* genes.[13,14] Cnidarian Hox-like genes for which the orthology relationships with bilaterian Hox genes are uncertain or debated (see Schierwater's chapter, this book) can be added.

The Hox Genes' Explosion

Although homeobox genes are present in a wide variety of eukaryotes, including plants and fungi, the Hox-extended family is present in the Animalia kingdom only. They are absent in the Porifera (sponges),[15] as evidenced by the deciphering of the complete genome of *Amphimedon queenslandica*, as well as intensive searches in other species.[16,17] In Ctenophores, up to now, no Hox-like gene has been found.[18] A single member of the Hox-extended family has been found in Placozoa.[19] In cnidarians, several members are found, several independent duplications have occurred from an estimated ancestral number of two to three genes depending on the authors (see Schierwater's chapter, this book).

The situation changes dramatically in the Bilateria. In an extensive search, de Rosa et al retrieved a number of Hox genes in a variety of bilaterian phyla, allowing the inference of a repertoire composed of seven to nine Hox genes in their common ancestor.[20] Hence, there has been a substantial increase in the number of Hox genes concomitant with the so-called 'Cambrian explosion', the sudden appearance of most extant bilaterian phyla.[21] In addition, the grouping and synteny of paralogous Hox genes in complexes in chordates,[22] hexapods[23] and annelids,[24] representatives of the three bilaterian super-phyla,[25] strongly suggest that the bilaterian ancestor[a] possessed a similar colinear Hox complex, despite further derived reorganization.[26] Another remarkable feature of bilaterian Hox genes is that they are primarily expressed in broad domains along the A-P axis,[27] whereas, despite some previous claims,[28] cnidarian Hox-like genes show more variable tissue- and species-specific expression patterns[29-32] (see Schierwater's chapter, this book).

What Is the Function of a Gene?

The founders of modern genetics, Thomas Hunt Morgan[33] and his colleagues let the issue of the function of genes aside.[b] The link between genes and metabolism was first raised by Garrod with his studies of hereditary diseases[34] and then approached experimentally by Ephrussi and Beadle[35] and Beadle and Tatum[36] leading to the famous one gene—one enzyme hypothesis.[37] With the advent of molecular genetics, the discovery of DNA as the molecular vector of genetic information,[38] of the structure of the DNA molecule[39] and the hypothesis and deciphering of the genetic code, a gene was taken as the DNA segment encoding a polypeptide, in agreement with the functional *cis-trans* test devised by Benzer.[40] Soon after, with the discovery of genetic regulation of the expression of bacterial genes, Jacob and Monod[41] introduced the distinction between 'structural genes' encoding proteins and "*other types of genetic determinants fulfilling specific function in control mechanisms*", including *cis*-acting genetic elements. It appears that in most eukaryotes, the latter largely exceed in sequence length that of coding sequences. *Cis*-acting sequences, critical for the in vivo function of a gene, may be located far from the coding sequence (see discussion about Hox genes in Spitz' chapter, this book). Can then the gene be reduced to its coding sequence, or even to the transcription unit? Alternatively, do we need to consider as the functional gene unit the whole sequence required to restore the wild-type

[a] In the present text, the 'bilaterian ancestor' means the common ancestor of protostomes and deuterostomes.

[b] T.H. Morgan (1928) The theory of the gene, chap II, p. 26: "The theory of the gene, as here formulated, states nothing with respect to the way in which the genes are connected with the end-product or character. [...] The sorting out of characters in successive generations can be explained at present without references to the way in which the gene affects the developmental process".

phenotype when introduced in null mutant by transgenesis? (see discussion about Hox genes in Maeda and Karch's chapter, this book). At the organism's level, the function of a gene can be revealed by the phenotype of gain- and loss-of-function mutants. As a phenotype, this functional definition is dependent on the context, including the genetic background and environmental factors. Hence, we have to precise at which level we consider the function of genes.

Hox Genes' Function at the Molecular and Cellular Levels

Soon after its discovery, it was shown that the homeodomain behaves as a DNA-binding domain.[42,43] In addition, McGinnis and colleagues showed that the Hox proteins have a transcriptional activity, either repressive or positive, alone or in association with cofactors[44] (see chapter by Merabet et al, this book). This identifies the Hox proteins as transcription factors. At the organism level, Garcia-Bellido[45] distinguished two types of developmental genes, selector and realizator ones. He allocated the Hox genes to the selector category. The molecular function of Hox genes, as encoding transcription factors, fits well with the predicted biological function as selectors. Another, more unexpected, function of homeodomain proteins has been evidenced more recently: they are able to cross cellular and nuclear membranes of neuronal cells, making it possible that they possess cell-communication activity[46] (see chapter by Merabet et al, this book).

Hox Genes and Homeosis

The word 'homeosis' was coined by Bateson.[47] In his book entitled "*Materials for the study of variation*", where he reported numerous examples of variations observed in the field in animals and plants, he wrote: "*For the word* 'Metamorphy' *I therefore propose to substitute the term* 'Homeosis', *which is also more correct; for the essential phenomenon is not that there has merely been a change, but that something has changed into the likeness of something else*". In brief, a homeotic transformation is not any monstrosity, but a change of a part of the organism into another, still recognizable. A more restrictive definition, applying to bilaterian animals, would be that homeosis is a transformation of any part of the body, a cell, tissue, organ, segment, into a corresponding part along the anterior to posterior (A-P) axis. Bateson gave such examples as the change of an antenna into a leg-like appendage, bearing a claw, in a hymenopteran insect and a change in vertebrae in a frog.

Soon after the launching of *Drosophila* genetics, Thomas H. Morgan and colleagues isolated homeotic mutants, but it was Richard Goldschmidt[48] who drew attention on their possible importance in evolution. The extensive search for homeotic genes in *Drosophila melanogaster* began in the mid-20th century with the works of Ed Lewis and Thom Kaufman (see above). Both loss- and gain-of-function mutations of these genes lead to homeotic transformation. Cloning of these Hox homeotic genes allowed studying their expression profiles, showing that they are expressed in overlapping broad regions spanning the A-P axis. Exceptions deal with four of them: the *zen*, *zen2* and *bicoid* genes, homologous to the *Hox3* paralogy group and *fushi tarazu* (*ftz*), homologous to the *Hox6* group.[49] Comparisons with other arthropod species have shown that the three *Drosophila Hox3* paralogs are issued from an ancient *Hox3* gene through recent duplication events and then they derived in both sequence and function (see Schmitt-Ott's chapter, this book). In chelicerate and myriapod species, *Hox3/zen* and *ftz* homologs[50-52] are expressed in a broad domain along the A-P axis, suggesting a primitive homeotic function in the arthropod ancestor.

Soon after the discovery of Hox genes in vertebrates, it appeared that their mutations led to homeotic transformations in the mouse.[9] This led to the hypothesis that the primitive function of Hox genes in the bilaterian ancestor was homeotic.

Homeosis as a Differential Function

From the link between mutations and homeosis, it was inferred that the function of Hox genes was to bring "identity" to various domains along the A-P bilaterian axis. But what does that mean exactly? In the fly, the 'identity' of a segment is defined by the type of its appendages: the first thoracic segment bears legs but no wings, the second legs and wings, the third legs and halteres, abdominal segments bear no legs and the posterior segments bear anal and genital appendages.

Indeed, the Hox genes determine which type of appendages is present in which segment. Similarly, in vertebrates, the Hox genes determine which type of vertebrae is made along the spinal cord, e.g., cervical, thoracic, lumbar, sacral. All along, there are vertebrae. Thus the 'identity' of a region of the A-P axis is brought by the morphological *differences* between vertebrae.

Although originally linked to developmental genetic compartments,[45] the concept of selector genes has been taken in a broader meaning, as genes specifying cell, tissue, organ morphogenesis.[53] It was thought that Hox genes were such organ-specifying genes. As an example, in flies, the 'identity' of the three thoracic segments is determined by the activity of the Hox genes *Sex combs reduced* (*Scr*), *Antennapedia* (*Antp*) and *Ultrabithorax* (*Ubx*), respectively. Hence, it was thought that *Antp* was involved in the formation of wings in the second thoracic segments. But this expectation was deceived. On the contrary, in the absence of *Antp* function, embryonic wing primordia are formed, rather, the other Hox genes, *Scr*, *Ubx* and *abdominal-A* (*abd-A*) repress wing formation in their respective domains.[54] In coleopterans, the second thoracic segment bears elytra, modified sclerotized wings used as a protective shield and the flight wings are the dorsal appendages of the third thoracic segment. In the beetle *Tribolium*, loss of function of the *Ubx* homolog leads to homeotic transformation of wings into elytra, which are a supposedly more derived form.[55] From these data, it can be concluded that the Hox genes are not involved as selectors of a given morphogenetic pathway for building a wing, a haltere or an elytron, but rather to set up the differences between the three thoracic segments, whatever the specific organogenesis could be.[56]

This does not exclude that in some cases Hox genes act directly on realizator genes to select organogenetic programmes.[57] However, these organs, such as the posterior spiracles of *Drosophila* larvae and the mammalian prostate, are obviously derived structures; hence, it can be inferred that Hox genes have been recruited for this direct morphogenetic function late in animal evolution.

Hox Genes as 'Meta-Selector' Genes

Thus, rather than 'selector' genes, Hox genes may be better viewed as 'meta-selectors', that is, selectors acting upstream of selector genes, not to perform any differentiating programme, e.g., muscular or haematopoietic, any organogenetic programme, e.g., heart or gut, but only to specify differences between regions along the A-P axis, independently of the type of cell, tissue or organ they may comprise.

This way of looking at the Hox genes' function is in agreement with the fact that among the known direct targets of the Hox genes, only few are 'realizator' genes, involved in cellular functions and most of them are themselves developmental genes, encoding transcription factors or cell-signalling proteins.[58]

The Hox Specificity Paradox

To ensure their homeotic function, the Hox genes were supposed to specifically control different morphogenetic programmes. Then it soon appeared that all Hox homeodomains recognize a short sequence, based on a core TAAT, with little discrimination. This has been called the "Hox specificity paradox". A part of Hox-proteins' specificity is due to differences in the Hox homeodomain itself, particularly in its N-terminal arm. Another part can be accounted for by interaction with cofactors, such as exd/PBX (see chapter by Merabet et al, this book). Viewing the Hox genes as 'meta-selectors' accounts for the loose specificity of the Hox proteins. Indeed, Hox activity reigns on various domains on the A-P axis of the trunk in bilaterian animals, almost exclusively on ectodermal and mesodermal derivatives. All these regions contain the same tissues, muscles, nerve cells, epidermis. The difference between them lies on subtle differences in cell and tissue relationships. It is then sensible that the Hox genes targets would always be the same along the A-P axis, such as developmental genes, selector and cell-signalling genes and even realizator genes, involved in cellular functions. The morphological differences specified by the Hox genes along the A-P axis would thus rely not on qualitative differences between the targets, but on subtle differences on the way as the *same* targets are activated or repressed and the relative timing of this control. A good example of the latter is given by *Ubx* activity in flies. In *Drosophila*, *Ubx* is known to repress *Distal less* (*Dll*), a gene required for the formation of appendages.[59] This is one of the best-studied targets of a Hox gene. *Dll* repression in abdominal segments by the Hox genes *Ubx* and *abdA* is causative of the lack of legs in the fly's abdomen. Still, a pair of legs is present in the third thoracic segment, despite *Ubx* expression. Lack of *Dll* repression in this segment is due to a delay in *Ubx* expression in the third thoracic vs expression in the first abdominal segment.[60]

Summarizing, the 'Hox specificity paradox' may be restated as follows: the various Hox genes do not act on different targets, but on the same targets, albeit differently. The issue is thus not to determine which gene is the target of which Hox protein, but how, where (precisely at the cellular level) and when, which Hox protein acts on a target.

Posterior Prevalence

A part of the answer may be given by interactions between Hox genes themselves. There are examples of transcriptional control of Hox genes by Hox products, including both positive and negative auto-regulation and regulation of a Hox gene by other Hox proteins. Nevertheless, the main interaction between Hox products is not at the transcriptional level, but between Hox proteins themselves. Indeed, it has been shown that when two Hox proteins are present in the same cell, the most 'posterior' one imposes its activity. This phenomenon has been called 'phenotypic suppression' in *Drosophila* and 'posterior prevalence' in vertebrates. It has been conserved between bilaterians throughout evolution.[61,62] Although the mechanism underlying posterior prevalence is not clearly elucidated yet, it is a posttranscriptional phenomenon. It may depend on specific differences in amino-acid sequence among the Hox paralogs,[63] possibly involving residues outside the homeodomain.[64] Noteworthy, in this respect, the length of the linker region between the two conserved regions of Hox proteins, the hexapeptide and the homeodomain, gradually decreases from the most anterior to the most posterior paralog groups[65] (see chapter by Merabet et al, this book). Recently, a role of miRNAs located within the Hox complexes has been suggested, which could be traced back to the bilaterian ancestor.[66] As underlined by Hombria and Lovegrove,[57] the differential activity of the Hox genes, i.e., their homeotic function, relies mostly on posterior prevalence.

An Evolutionary Paradox: Morphological Differentiation and the Hox Repertoire

It was previously thought that the Hox repertoire should increase with the increasing morphological complexity in the various animal lineages.[1,67] This prediction has not been vindicated. On the contrary, in the panarthropod lineages, animals with homonomous segmentation, such as onychophorans[68] and myriapods[52] have exactly the same Hox genes' complement as animals with diversified trunk morphology, such as chelicerates, crustaceans and hexapods. Similarly, in chordates, the amphioxus possesses up to 15 Hox genes,[69] the coelacanth and a shark 14,[11] whereas the complement of a vertebrate Hox cluster is 13,[9,10] suggesting a loss of Hox paralogs in teleosts and in tetrapods, despite increased diversification of the body axis. In the third bilaterian super-phylum, the polychaete annelids *Capitella*[24] and Nereids[20,70] display the primitive lophotrochozoan number of 11 Hox genes, without presenting any obvious morphological differentiation between trunk segments.

Along the usual definition of homeosis, as the change of a part of the body into the likeness of something else, overt morphology is concerned. We have seen above that the function of the Hox genes is to make differences between regions along the A-P axis, whatever the morphogenetic programme specific to these regions be. What is then the function of the Hox genes when there is no morphological difference?

Hox and Neuronal Homeosis

I have previously suggested that the primitive function of the Hox genes is to determine neuronal differentiation along the A-P axis.[71] Hox genes' expression in the central nervous system has been well documented by numerous publications in vertebrates[72] and in Drosophila[73] (see chapter by Reichert and Bello, this book). This extends to members of the third super-phylum of bilaterian animals, the Lophotrochozoa, such as leeches,[74] polychaete annelids[24,70,75] and molluscs.[76] In some organisms, Hox genes are expressed almost exclusively in the CNS; this is the case in leeches[74] and in the amphioxus.[77] Moreover, A-P regionalized expression is observed in central nervous systems presenting such different morphologies as the ganglionic nerve cord of annelids and of arthropods, the dorsal neural tube of chordates (see above), the cycloneuralian system of nematodes,[78,79] the sub-epidermal diffuse nervous system of hemichordates.[80,81] Thus, again, the differential Hox pattern along the A-P axis of the Hox genes is conserved irrespective of the actual morphology of the CNS.

One may ask whether this differential pattern actually corresponds to a differential function. In other terms, do mutations in Hox genes lead to what can be called "neuronal homeosis", i.e., transformation of the neuronal architecture of a specific body region to that of another region of the A-P axis? The answer is yes, indeed. As soon as in the 1980', the nervous network of the *Drosophila bithorax* mutant was examined both on anatomical and on functional levels.[82-85] More recently, a new, previously unexpected, function of the Hox genes in sculpting the nervous network of *Drosophila* was revealed: Hox genes control differential apoptosis of neuronal precursors, pioneer and differentiated neurons[86-89] (see chapter by Reichert and Bello, this book).

Similarly, neuronal homeosis is observed in mice, both in gain- and loss-of-function mutants and through RNA interference in Hox genes.[90-94] Hox regulation of neural apoptosis is also observed.[95-97] In the third bilaterian super-phylum functional genetic experiments are scarce. However, in leeches, where the location of precise neurons is specific and reproducible, ectopic expression by injecting mRNA of the Hox gene *Lox1* was followed by a change of the electrical properties of specific neurons.[98]

So, Hox genes design the nervous system altogether in specifying the location of particular neuroblasts, tracing the route of axons and eliminating supplementary nerve cells by inducing apoptosis. The final result is differentiation of the CNS and nerve net along the A-P axis. This function of the Hox genes can likely be traced back to the bilaterian ancestor.

Morphological Homeosis as a Derived Property

We have previously seen that the number and differential expression of Hox genes does not depend on the actual morphological differentiation of the body along the A-P axis. Indeed, animals with more uniform or more diversified morphologies, sharing the same Hox gene complement and homologous Hox patterns, are found in various phyla. This is a strong argument to think that homeosis, in the usual meaning of *morphological* differentiation is a derived property. On the other side, neuronal homeosis could well be ancestral in bilaterians. From the presence of a device generating neuronal specificity along the A-P axis, the same device, the 'Hox system' could have been recruited multiple times during bilaterian evolution to generate morphological diversity.

Similarly, the 'Hox system' has been recruited to generate diversity in secondary axes, such as the paired fins and tetrapod limbs of vertebrates.[99,100]

Why Does the 'Hox System' Make Sense?

Recalling Dobzhansky's famous aphorism "*Nothing in biology makes sense except in the light of evolution*", can we try to answer the question "*Why the Hox system*"?

First, although Hox-like genes precede the emergence of the Bilateria, the 'Hox system', as a complete set of about 9 to 10 differentiated genes with differentiated expression domains do not (see chapter by Scheirwater, this volume).

Second, a CNS comprising a dorsal anterior 'brain' and a trunk nerve cord is a synapomorphy of the Bilateria. There is accumulating evidence for a common genetic patterning of the bilaterian CNS, irrespective of its morphological diversity[101-103] (see chapter by Reichert and Bello, this book).

In bilaterian animals, the trunk nerve cord and the 'Hox system' are both oriented along the A-P axis. It is sensible to think that, to ensure correct movements of the animal, a correspondence is needed along the A-P axis between the nerve cord and the rest of the body. In other terms, the CNS must 'know' which part of the body motor nerves make synapses with and which part of the body sensory neurons project from. Without such correspondence, the behaviour of the animal would be uncoordinated. This requires positional information along the A-P axis of the CNS, in both motor and sensory directions. This information would be translated into regional-specific nervous architectures, even in animals where there is no overt morphological differentiation. *I postulate the Hox system has been the evolutionary innovation that fulfilled this need.*

Conclusion

In conclusion, although the Hox genes predate the bilaterians radiation, the 'Hox system', defined as a complex of Hox genes acting in coordination, is specific to the Bilateria. The Hox genes have been exapted[104] from the Hox-like cnidarian Hox genes of unknown function to build the Hox system. In the bilaterians, the Hox genes appear as 'meta-selector' genes, controlling both realizator and selector genes. The concept of 'meta-selectors' resolves the 'Hox paradox' on the alleged contradiction between the loose specificity of the Hox homeodomains on their DNA targets and the discriminating function of the Hox genes at the organismal level, because all Hox proteins act on the same targets, with only subtle differences on their time and location of activity and interactions with other partners.

The function of the bilaterian Hox genes is not merely to determine a precise morphology in various parts of the body on the A-P axis, but to ensure that they would be different from each other. The need for such differences is engraved within the construction of the bilaterian body plan with a central nervous system built along the A-P axis. As all bilaterian animals are motile at least during some part of their life cycle, the need for mobility requires a correspondence between the parts of the body and sections of the CNS along the A-P axis, unless the animal would be uncoordinated. The function of the 'Hox system' in bilaterians would thus be to ensure this correspondence by drawing the nerve net properly in both motor and sensory directions. Thus the primary function of the Hox genes in both evolution and development in the Bilateria is neuronal, their so-called 'homeotic' function, in the meaning of determining morphological 'identities' of the various parts of the body along the A-P axis being derived.

This hypothesis accounts for all present observations. It is also testable through the study of Hox genes in animals presenting homonomous morphology, such as annelids, remipede crustaceans, myriapods and cephalochordates: their expression should be regionalized (as it is the case in myriapods and in the amphioxus) and their loss of function by transgenesis, RNA interference or morpholinos should lead to neuronal homeosis.

References

1. Lewis EB. A gene complex controlling segmentation in Drosophila. Nature 1978; 276:565-570.
2. Kaufman TC, Lewis R, Wakimoto B. Cytogenetic Analysis of Chromosome 3 in Drosophila melanogaster: The homoeotic gene complex in polytene chromosome interval 84A-B. Genetics 1980; 94:115-133.
3. Scott MP, Weiner AJ. Structural relationships among genes that control development: sequence homology between the Antennapedia, Ultrabithorax and fushi tarazu loci in Drosophila. Proc Natl Acad Sci USA 1984; 81:4115-4119.
4. McGinnis W, Levine MS, Hafen E et al. A conserved DNA sequence in homoeotic genes of the Drosophila Antennapedia and Bithorax Complexes. Nature 1984a; 308:428-433.
5. McGinnis W, Garber RL, Wirz J et al. A homologous protein-coding sequence in Drosophila homeotic genes and its conservation in other metazoans. Cell 1984b; 37:403-408.
6. Duboule D. Guidebook to the homeobox genes. Oxford: Oxford University Press, 1994:284.
7. Derelle R, Lopez P, Le Guyader H et al. Homeodomain proteins belong to the ancestral molecular toolkit of eukaryotes. Evol Dev 2007; 9:212-219.
8. Zhong YF, Butts T, Holland PW. HomeoDB: a database of homeobox gene diversity. Evol Dev 2008; 10:516-518.

9. McGinnis W, Krumlauf R. Homeobox genes and axial patterning. Cell 1992; 68:283-302.
10. Scott MP. Vertebrate homeobox gene nomenclature. Cell 1992; 71:551-553.
11. Powers TP, Amemiya CT. Evidence for a Hox14 paralog group in vertebrates. Curr Biol 2004; 14:R183-R184.
12. Brooke NM, Garcia-Fernandez J, Holland PWH. The ParaHox gene cluster is an evolutionary sister of the Hox gene cluster. Nature 1998; 392:920-922.
13. Bharathan G, Janssen BJ, Kellogg EA et al. Did homeodomain proteins duplicate before the origin of angiosperms, fungi and metazoa? Proc Natl Acad Sci USA 1997; 94:13749-13753.
14. Pollard SL, Holland PW. Evidence for 14 homeobox gene clusters in human genome ancestry. Curr Biol 2000; 10:1059-1062.
15. Larroux C, Fahey B, Degnan SM et al. The NK homeobox gene cluster predates the origin of Hox genes. Curr Biol 2007; 17:706-10. Epub 2007 Mar 22.
16. Seimiya M, Ishiguro H, Mura K et al. Homeobox-containing genes in the most primitive metazoa, the sponges. Eur J Biochem 1994; 221:219-225.
17. Manuel M, Le Parco Y. Homeobox gene diversification in the calcareous sponge, Sycon raphanus. Mol Phylogenet Evol 2000; 17:97-107.
18. Pang K, Martindale MQ. Developmental expression of homeobox genes in the ctenophore Mnemiopsis leidyi. Dev Genes Evol 2008; 218:307-319.
19. Schierwater B, Kuhn K. Homology of Hox genes and the zootype concept in early metazoan evolution. Mol Phylogenet Evol 1998; 9:375-381.
20. de Rosa R, Grenier J, Andreeva T et al. Hox genes in brachiopods and priapulids and protostome evolution. Nature 1999; 399:772-776.
21. Deutsch JS, Lopez P. Are transposition events at the origin of bilaterian Hox complexes? In: Minelli A, Fusco G, eds. Evolving Pathways. Cambridge (UK): Cambridge University Press, 2008:239-260.
22. Garcia-Fernandez J, Holland PWH. Archetypal organization of the amphioxus Hox gene cluster. Nature 1994; 370:563-566.
23. Richards S, Gibbs RA, Weinstock GM et al. The genome of the model beetle and pest Tribolium castaneum. Nature 2008; 452:949-955.
24. Fröbius AC, Matus DQ, Seaver EC. Genomic organization and expression demonstrate spatial and temporal Hox gene colinearity in the lophotrochozoan Capitella sp. I. PLoS ONE 2008; 3:e4004.
25. Adoutte A, Balavoine G, Lartillot N et al. The new animal phylogeny: reliability and implications. Proc Natl Acad Sci USA 2000; 97:4453-4456.
26. Duboule D. The rise and fall of Hox gene clusters. Development 2007; 134:2549-2560.
27. Slack JMW, Holland PWH, Graham CF. The zootype and the phylotypic stage. Nature 1993; 361:490-492.
28. Finnerty JR, Pang K, Burton P et al. Origins of bilateral symmetry: Hox and dpp expression in a sea anemone. Science 2004; 304:1335-1337.
29. Masuda-Nakagawa LM, Groer H, Aerne BL et al. The HOX-like gene Cnox2-Pc is expressed at the anterior region in all life cycle stages of the jellyfish Podocoryne carnea. Dev Genes Evol 2000; 210:151-156.
30. Kamm K, Schierwater B, Jakob W et al. Axial patterning and diversification in the Cnidaria predate the Hox system. Curr Biol 2006; 16:920-926.
31. Ryan JF, Mazza ME, Pang K et al. Pre-bilaterian origins of the Hox cluster and the Hox code: Evidence from the sea anemone Nematostella vectensis. PLoS ONE 2007; 2:e153.
32. Chiori R, Jager M, Denker E et al. Are Hox genes ancestrally involved in axial patterning? Evidence from the hydrozoan Clytia hemisphaerica (Cnidaria). PLoS ONE 2009; 4:e4231.
33. Morgan TH. The theory of the gene. New Haven: Yale University Press, 1928:358.
34. Garrod A. The incidence of alkaptonuria: A study in chemical individuality. Lancet 1902; ii:1616-1620.
35. Ephrussi B. Aspects of the physiology of gene action. Amer Nat 1938; 72:5-23.
36. Beadle GW, Tatum EL. Genetic control of biochemical reactions in Neurospora. Proc Nat Acad Sci USA 1941; 27:499-506.
37. Beadle GW. Biochemical genetics: Some recollections. In: Cairns J, Stent GS, Watson JD, eds. Phage and the origin of molecular biology. Cold Spring Harbor: Cold Spring Harbor Laboratory 1966:23-32.
38. Avery OT, MacLeod CM, McCarty M. Studies on the chemical nature of the substance inducing transformation of pneumococcal types. J Exp Med 1944; 79:137-158.
39. Watson JD, Crick FHC. Molecular structure of nucleic acids. A structure of deoxyribose nucleic acid. Nature 1953; 171:737-738.

40. Benzer S. Fine structure of a genetic region in bacteriophage. Proc Nat Acad Sci USA 1955; 41:344-354.
41. Jacob F, Monod J. Genetic regulatory mechanisms in the synthesis of proteins. J Mol Biol 1961; 3:318-356.
42. Qian YQ, Billeter M, Otting G et al. The structure of the Antennapedia homeodomain determined by NMR spectroscopy in solution: comparison with prokaryotic repressors. Cell 1989; 59:573-580.
43. Qian YQ, Otting G, Billeter M et al. Nuclear magnetic resonance spectroscopy of a DNA complex with the uniformly ^{13}C-labeled Antennapedia homeodomain and structure determination of the DNA-bound homeodomain. J Mol Biol 1993; 234:1070-1083.
44. Li X, McGinnis W. Activity regulation of Hox proteins, a mechanism for altering functional specificity in development and evolution. Proc Natl Acad Sci USA 1999; 96:6802-6807.
45. Garcia-Bellido A. Genetic control of wing disc development in Drosophila. In: Brenner S, ed. Cell Patterning. Amsterdam: Elsevier, 1975:161-182.
46. Prochiantz A, Joliot A. Can transcription factors function as cell-cell signalling molecules? Nat Rev Mol Cell Biol 2003; 4:814-819.
47. Bateson W. Materials for the study of variation, treated with especial regard to discontinuity in the origin of species. London: MacMillan & Co, 1894;
48. Goldschmidt RB. The material basis of evolution. New Haven: Yale University Press, 1940 (1982):436.
49. Telford MJ. Evidence for the derivation of the Drosophila fushi tarazu gene from a Hox gene orthologous to lophotrochozoan Lox5. Curr Biol 2000; 10:349-352.
50. Telford MJ, Thomas RH. Of mites and zen: expression studies in a chelicerate arthropod confirm zen is a divergent Hox gene. Dev Genes Evol 1998; 208:591-594.
51. Damen WG, Tautz D. A Hox class 3 orthologue from the spider Cupiennius salei is expressed in a Hox-gene-like fashion. Dev Genes Evol 1998; 208:586-590.
52. Hughes CL, Kaufman TC. Exploring the myriapod body plan: expression patterns of the ten Hox genes in a centipede. Development 2002; 129:1225-1238.
53. Mann RS, Carroll SB. Molecular mechanisms of selector gene function and evolution. Curr Opin Genet Dev 2002; 12:592-600.
54. Carroll SB, Weatherbee SD, Langeland JA. Homeotic genes and the regulation and evolution of insect wing number. Nature 1995; 375:58-61.
55. Tomoyasu Y, Wheeler SR, Denell RE. Ultrabithorax is required for membranous wing identity in the beetle Tribolium castaneum. Nature 2005; 433:643-647.
56. Deutsch J. Hox and wings. Bioessays 2005; 27:673-675.
57. Hombria JC, Lovegrove B. Beyond homeosis—HOX function in morphogenesis and organogenesis. Differentiation 2003; 71:461-476.
58. Graba Y, Aragnol D, Pradel J. Drosophila Hox complex downstream targets and the function of homeotic genes. Bioessays 1997; 19:379-388.
59. Vachon G, Cohen B, Pfeifle C et al. Homeotic genes of the Bithorax complex repress limb development in the abdomen of the Drosophila embryo through the target gene Distal-less. Cell 1992; 71:437-450.
60. Castelli-Gair J, Akam M. How the Hox gene Ultrabithorax specifies two different segments: the significance of spatial and temporal regulation within metameres. Development 1995; 121:2973-2982.
61. Bachiller D, Macias A, Duboule D et al. Conservation of a functional hierarchy between mammalian and insect Hox/HOM genes. EMBO J 1994; 13:1930-1941.
62. Duboule D, Morata G. Colinearity and functional hierarchy among genes of the homeotic complexes. Trends Genet 1994; 10:358-364.
63. Merabet S, Hudry B, Saadaoui M et al. Classification of sequence signatures: a guide to Hox protein function. Bioessays 2009; 31:500-511.
64. Williams ME, Lehoczky JA, Innis JW. A group 13 homeodomain is neither necessary nor sufficient for posterior prevalence in the mouse limb. Dev Biol 2006; 297:493-507.
65. In der Rieden PM, Mainguy G, Woltering JM et al. Homeodomain to hexapeptide or PBC-interaction-domain distance: size apparently matters. Trends Genet 2004; 20:76-79.
66. Yekta S, Tabin CJ, Bartel DP. MicroRNAs in the Hox network: an apparent link to posterior prevalence. Nat Rev Genet 2008; 9:789-796.
67. Akam M, Dawson I, Tear G. Homeotic genes and the control of segment diversity. Development Supp 1988; 104 Supp.:123-133.
68. Grenier JK, Garber TL, Warren R et al. Evolution of the entire arthropod Hox gene set predated the origin and radiation of the onychophoran/arthropod clade. Curr Biol 1997; 7:547-553.

69. Amemiya CT, Prohaska SJ, Hill-Force A et al. The amphioxus Hox cluster: characterization, comparative genomics and evolution. J Exp Zoolog B Mol Dev Evol 2008; 310:465-477.
70. Kulakova M, Bakalenko N, Novikova E et al. Hox gene expression in larval development of the polychaetes Nereis virens and Platynereis dumerilii (Annelida, Lophotrochozoa). Dev Genes Evol 2007; 217:39-54.
71. Deutsch J, Le Guyader H. The neuronal zootype: a hypothesis. Compt R Acad Sci III (Paris) 1998; 321:713-719.
72. Krumlauf R, Marshall H, Studer M et al. Hox homeobox genes and regionalisation of the nervous system. J Neurobiol 1993; 24:1328-1340.
73. Technau GM, Berger C, Urbach R. Generation of cell diversity and segmental pattern in the embryonic central nervous system of Drosophila. Dev Dyn 2006; 235:861-869.
74. Kourakis MJ, Master VA, Lokhorst DK et al. Conserved anterior boundaries of Hox gene expression in the central nervous system of the leech Helobdella. Dev Biol 1997; 190:284-300.
75. Irvine SQ, Martindale MQ. Expression patterns of anterior Hox genes in the polychaete Chaetopterus: correlation with morphological boundaries. Dev Biol 2000; 217:333-351.
76. Hinman VF, O'Brien EK, Richards GS et al. Expression of anterior Hox genes during larval development of the gastropod Haliotis asinina. Evol Dev 2003; 5:508-521.
77. Wada H, Garcia-Fernandez J, Holland PW. Colinear and segmental expression of amphioxus Hox genes. Dev Biol 1999; 213:131-141.
78. Salser SJ, Loer CM, Kenyon C. Multiple HOM-C gene interactions specify cell fates in the nematode central nervous system. Genes Dev 1993; 7:1714-1724.
79. Hunter CP, Kenyon C. Specification of anteroposterior cell fates in Caenorhabditis elegans by Drosophila Hox proteins. Nature 1995; 377:229-232.
80. Lowe CJ, Wu M, Salic A et al. Anteroposterior patterning in hemichordates and the origins of the chordate nervous system. Cell 2003; 113:853-865.
81. Aronowicz J, Lowe CJ. Hox gene expression in the hemichordate Saccoglossus kowalevskii and the evolution of deuterostome nervous systems. Integr Comp Biol 2006; 46:890-901.
82. Palka J, Lawrence PA, Hart HS. Neural projection patterns from homeotic tissue of Drosophila studied in bithorax mutants and mosaics. Dev Biol 1979; 69:549-575.
83. Ghysen A. The projection of sensory neurons in the central nervous system of Drosophila: choice of the appropriate pathway. Dev Biol 1980; 78:521-541.
84. Burt R, Palka J. The central projections of mesothoracic sensory neurons in wild-type Drosophila and bithorax mutants. Dev Biol 1982; 90:99-109.
85. Schneiderman AM, Tao ML, Wyman RJ. Duplication of the escape-response neural pathway by mutation of the Bithorax-Complex. Dev Biol 1993; 157:455-473.
86. Bello BC, Hirth F, Gould AP. A pulse of the Drosophila Hox protein Abdominal-A schedules the end of neural proliferation via neuroblast apoptosis. Neuron 2003; 37:209-219.
87. Miguel-Aliaga I, Thor S. Segment-specific prevention of pioneer neuron apoptosis by cell-autonomous, postmitotic Hox gene activity. Development 2004; 131:6093-6105.
88. Rogulja-Ortmann A, Luer K, Seibert J et al. Programmed cell death in the embryonic central nervous system of Drosophila melanogaster. Development 2007; 134:105-116.
89. Rogulja-Ortmann A, Renner S, Technau GM. Antagonistic roles for Ultrabithorax and Antennapedia in regulating segment-specific apoptosis of differentiated motoneurons in the Drosophila embryonic central nervous system. Development 2008; 135:3435-3445.
90. Zhang M, Kim HJ, Marshall H et al. Ectopic Hoxa-1 induces rhombomere transformation in mouse hindbrain. Development 1994; 120:2431-2442.
91. Studer M, Lumsden A, Ariza-McNaughton L et al. Altered segmental identity and abnormal migration of motor neurons in mice lacking Hoxb-1. Nature 1996; 384:630-634.
92. Gavalas A, Ruhrberg C, Livet J et al. Neuronal defects in the hindbrain of Hoxa1, Hoxb1 and Hoxb2 mutants reflect regulatory interactions among these Hox genes. Development 2003; 130:5663-5679.
93. Lin AW, Carpenter EM. Hoxa10 and Hoxd10 coordinately regulate lumbar motor neuron patterning. J Neurobiol 2003; 56:328-337.
94. Dasen JS, Tice BC, Brenner-Morton S et al. A Hox regulatory network establishes motor neuron pool identity and target-muscle connectivity. Cell 2005; 123:477-491.
95. Tiret L, Le Mouellic H, Maury M et al. Increased apoptosis of motoneurons and altered somatotopic maps in the brachial spinal cord of Hoxc-8-deficient mice. Development 1998; 125:279-291.
96. del Toro ED, Borday V, Davenne M et al. Generation of a novel functional neuronal circuit in Hoxa1 mutant mice. J Neurosci 2001; 21:5637-5642.
97. Economides KD, Zeltser L Capecchi MR. Hoxb13 mutations cause overgrowth of caudal spinal cord and tail vertebrae. Dev Biol 2003; 256:317-330.

98. Aisemberg GO, Gershon TR, Macagno ER. New electrical properties of neurons induced by a homeoprotein. J Neurobiol 1997; 33:11-17.
99. Shubin N, Tabin C, Carroll S. Fossils, genes and the evolution of animal limbs. Nature 1997; 388:639-648.
100. Tarchini B, Duboule D, Kmita M. Regulatory constraints in the evolution of the tetrapod limb anterior-posterior polarity. Nature 2006; 443:985-988.
101. Arendt D, Nübler-Jung K. Common ground plans in early brain development in mice and flies. BioEssays 1996; 18:255-259.
102. Reichert H, Simeone A. Developmental genetic evidence for a monophyletic origin of the bilaterian brain. Philos Trans R Soc Lond B Biol Sci 2001; 356:1533-1544.
103. Denes AS, Jekely G, Steinmetz PR et al. Molecular architecture of annelid nerve cord supports common origin of nervous system centralization in Bilateria. Cell 2007; 129:277-288.
104. Gould SJ, Vrba ES. Exaptation – a missing term in the science of form. Paleobiol 1982; 8:4-15.

Index

P

R

S

T

V

W

Z

GPSR Compliance

The European Union's (EU) General Product Safety Regulation (GPSR) is a set of rules that requires consumer products to be safe and our obligations to ensure this.

If you have any concerns about our products, you can contact us on ProductSafety@springernature.com

In case Publisher is established outside the EU, the EU authorized representative is:

Springer Nature Customer Service Center GmbH
Europaplatz 3
69115 Heidelberg, Germany

Batch number: 10762179

Printed by Printforce, the Netherlands